$37.95

WORLD INDUSTRY STUDIES 5

Electronics and Industrial Policy

WORLD INDUSTRY STUDIES

*Edited by Professor Ingo Walter,
Graduate School of Business Administration,
New York University*

1. COLLAPSE AND SURVIVAL: INDUSTRY STRATEGIES IN A CHANGING WORLD
   Robert H. Ballance and Stuart W. Sinclair

2. THE GLOBAL TEXTILE INDUSTRY
   Brian Toyne, Jeffrey S. Arpan, Andy H. Barnett, David A. Ricks and Terence A. Shimp

3. BLUE GOLD: THE POLITICAL ECONOMY OF NATURAL GAS
   J. D. Davis

4. POLITICS VS ECONOMICS IN WORLD STEEL TRADE
   Kent Jones

5. ELECTRONICS AND INDUSTRIAL POLICY
   The case of computer controlled lathes
   Staffan Jacobsson

Electronics and Industrial Policy
The case of computer controlled lathes

Staffan Jacobsson
University of Lund, Sweden

LONDON
ALLEN & UNWIN
Boston   Sydney

© Staffan Jacobsson, 1986
This book is copyright under the Berne Convention
No reproduction without permission. All rights reserved

Allen & Unwin (Publishers) Ltd,
40 Museum Street, London WC1A 1LU, UK

Allen & Unwin (Publishers) Ltd,
Park Lane, Hemel Hempstead, Herts HP2 4TE, UK

Allen & Unwin, Inc.,
8 Winchester Place, Winchester, Mass. 01890, USA

Allen & Unwin (Australia) Ltd,
8 Napier Street, North Sydney, NSW 2060, Australia

First published in 1986

British Library Cataloguing in Publication Data

Jacobsson, Staffan
  Electronics and industrial policy: the
  case of computer-controlled lathes.
1. Technological innovations – Government
policy 2. Industry and state
I. Title
338' 06   HD45
ISBN 0-04-338138-3

Library of Congress Cataloging-in-Publication Data

Jacobsson, Staffan.
  Electronics and industrial policy.
Bibliography: p.
Includes index.
1. Lathes – Numerical control. I. Title.
TJ1218.J24 1986   338.4'7621.942   86-7913
ISBN 0-04-338138-3 (alk. paper)

Set in 10 on 11 point Times by
Phoenix Photosetting, Chatham
and printed and bound in Great Britain by
Billing and Son Ltd, London and Worcester

Contents

	page
Preface	ix
Acknowledgements	xi
List of Abbreviations	xiii
List of Tables	xv
List of Figures	xx

1 Introduction — 1

2 The Technology and its Diffusion — 6
2.1 Introduction — 6
2.2 CNC lathe technology — 8
2.3 Choice of technique in turning — 12
2.4 The factor-saving bias of CNC lathes — 18
2.4.1 Some examples from Sweden — 18
2.4.2 The effects on the use of the various factors of production — 21
2.5 Conclusions — 28

3 Growth and Market Structure in the International CNC Lathe Industry — 31
3.1 Introduction — 31
3.2 A theoretical framework for analysing the industry — 37
3.2.1 The industry in a dynamic context — 43
3.3 The Japanese expansion in the CNC lathe industry — 47
3.3.1 A discussion of the factors behind the Japanese success — 52
3.4 The European response — 66
3.4.1 Overall cost leadership strategy — 67
3.4.2 Focus strategy — 68
3.4.3 Differentiation strategy — 71
3.5 A note on the US producers — 79
3.6 Concluding remarks on the strategies pursued by firms based in the OECD — 81
Notes — 86

CONTENTS

4	**Barriers to entry into the Overall Cost Leadership Strategy**	88
	4.1 Research and Development	88
	4.2 Procurement of components	91
	4.3 Manufacturing	94
	4.4 Marketing and after-sales services	95
	4.5 An attempt to specify the minimum efficient scale of production	98
	Notes	100
5	**The position of the NICs within the CNC Lathe Industry**	102
	5.1 The position of eight NIC-based firms within the low-performance strategy	105
	Notes	111
6	**The Case of Argentina**	113
	6.1 Growth and structure of the engineering industry	113
	6.2 The Argentinian machine tool industry	115
	6.3 Government policy	119
	6.4 The diffusion of CNC lathes in Argentina	121
	6.4.1 The historical rate of diffusion of CNC lathes in Argentina	122
	6.4.2 An estimate of the future rate of diffusion	125
	6.4.3 Conclusion	128
	6.5 The firm producing CNC lathes	128
	6.5.1 The full export oriented strategy	132
	6.5.2 The regional market strategy	139
	6.6 Summary and conclusions	141
	Notes	141
7	**The Case of Taiwan**	143
	7.1 Growth and structure of the engineering industry	143
	7.2 The Taiwanese machine tool industry	146
	7.3 Government policy	150
	7.4 The diffusion of CNC lathes in Taiwan	153
	7.5 The firms producing CNC lathes	154
	7.6 Government policy in the machine tool field	163
	7.6.1 Tariff policy	164
	7.6.2 Provision of risk capital	164
	7.6.3 The Export-Import Bank	165
	7.6.4 R&D on automation	166

CONTENTS

7.6.5	China External Trade Development Council	167
7.7	Evaluating the explicit governmental policy	168
7.7.1	Tariffs	168
7.7.2	The Export-Import Bank	169
7.7.3	Risk capital	170
7.7.4	R&D on automation	170
7.8	Summary and conclusions	172
	Note	173

8 The Case of Korea — 174
- 8.1 Growth and structure of the engineering industry — 174
- 8.2 The Korean machine tool industry — 177
- 8.3 Government policy — 177
- 8.3.1 The machine tool industry — 180
- 8.4 The Korean market for CNC lathes — 182
- 8.5 The firms producing CNC lathes — 184
- 8.6 Evaluating governmental policy — 189
- 8.7 Summary and conclusions — 190
- Notes — 192

9 Government Policy — 193
- 9.1 Introduction — 193
- 9.2 Survey of the main arguments — 194
- 9.2.1 The infant industry argument — 197
- 9.2.2 General versus selective state intervention — 197
- 9.2.3 The choice of policy instrument — 199
- 9.3 Government policy and industry performance in Argentina, Korea and Taiwan — 201
- 9.3.1 The benefits and costs of fostering the machine tool industry in the three NICs — 201
- 9.3.2 Implications for government policy — 208
- 9.4 Industry performance and implications for government policy in small developed countries — 221
- 9.4.1 The European industry — 221
- 9.4.2 Sweden and the UK — 226
- 9.5 Conclusions — 232
- Notes — 234

10 Summary — 235
- 10.1 The diffusion of CNC lathes — 235

10.2		The international CNC lathe producing industry	236
10.3		The NIC experience	237
10.4		Government policies	238

References   240

Index   249

To my family

Preface

There is a rapidly expanding literature on the economics of the so-called 'new technologies' – especially on those using microelectronic systems. Dr. Jacobsson's book deals with microelectronics-based innovation in machine tools: with the production and use of computer numerically controlled machine tools in the world economy and especially in the Third World. Jacobsson is mainly interested in the implications which CNC machine tools may be expected to have for users and producers in the Newly Industrialising Countries. He approaches this as a problem in applied economics and the book will have a primary interest for those economists whose concern is with the problems of industrialisation in developing countries. It will be particularly valuable to those who are preoccupied with the role of local capital goods manufacture and with the technological preconditions for this kind of production. Jacobsson is able to give detailed and specific arguments on these matters as far as CNC machine tools are concerned.

In my view, the book has a considerably wider interest and relevance than its specification may at first sight suggest. Jacobsson's achievement is not just that he has provided valuable and convincing quantitative arguments about policy in setting up production of CNC machine tools. In addition, he has set a new and much needed methodological standard for analysis of the impacts of 'new technologies' on the international economy.

The need for some tightening up of analytic standards is quickly apparent to anyone who is obliged to read the literature on the effects of 'new technologies' in the international economy. That literature has a number of very obvious deficiencies. First, much of it proceeds as if there were no preceding theoretical or empirical economic research on technology and technical change on which to draw. Implicitly authors seem to assume that 'old' analysis – for example on the economics of innovation, or on the behaviour of the firm, or on the economics of oligopolistic markets – is somehow unhelpful to our understanding of 'new' technologies. Second, although the literature is in some respects commendably empirical, the empiricism too often becomes anecdotal and arbitrary – mainly because of the lack of theoretical focus. Third, partly because of the first two deficiencies, virtually all the studies that are available are weak on questions of policy.

Jacobsson's study is exciting precisely because it rises above these limitations. He has successfully deployed relevant pieces of economic theory to provide a framework for his analysis. Chapter Two contains a discussion of such matters as the choice of technique in machine-tool technologies and the factor saving bias associated with CNC technology. It is one of very few analyses available which provides clear empirically informed results on these questions. Chapter Three gives an account of the operation of the highly imperfect world market for various categories of CNC lathes. It is well founded in fact and together with Chapter Four on barriers to entry in the international industry, provides an economically realistic framework for the subsequent discussion.

There follows a set of highly interesting chapters (Chapters Five to Eight), which analyse the CNC lathe industries in Argentina, Taiwan and South Korea. These are perhaps the most detailed and carefully observed studies available anywhere, of the adoption of a 'new technology' industry in Third World economies. Once again in these chapters, Jacobsson's empiricism is directed consistently to hypotheses which are sensibly derived from whatever useful theory is available. Finally, Chapter Nine is an analysis of the case for Government intervention in favour of a subsidised infant industry development of the CNC industry. It is one of few properly organised discussions of Third World industrial policy towards a 'new technology' industry. One does not have to agree all the way with Jacobsson's proposals to recognise that he has provided a convincing framework for decision.

Obviously there remains plenty of room for empirical study of new technology impacts in the Third World – and also of the particular implications of the CNC machine tools for developing capital goods sectors. I am sure that Staffan Jacobsson would agree with me that he has not had the last word on this interesting and important subject. What he has done, however, is to raise the standard of economic discussion on new technologies in an appreciable way. He has demonstrated that economic ideas and theories can be used creatively to direct empirical enquiry on these matters and in doing so he has made an original and much needed contribution.

Charles Cooper,
Maastricht and The Hague,
August 1986.

Acknowledgements

This book is based on my Doctor of Philosophy thesis in economics at the University of Sussex, which was accepted in 1985. My interest in the general field of technology and development was first awakened when I happened to visit the Research Policy Institute as an undergraduate at the University of Lund in Sweden. Professor Jon Sigurdson introduced me to the subject and encouraged me to spend some time at the University of Sussex because of the considerable expertise concentrated there in this field of study. Professor Charles Cooper cemented my interest in the field of the economics of technology and industrial development during my MA in Development Economics at the University of Sussex.

The writing of a book is a long and sometimes painful task. In the course of the work, I depended on a large number of people for support and for information. Intellectual, emotional and monetary debts have accumulated. It is not an easy task to try to provide a reasonably accurate picture of these debts; an attempt will however be made.

Lesley Palmer has given me unending moral support and encouragement. She has also shown a keen interest in the subject and is by now probably an expert in the field of study. Her hours spent on correcting my English are probably appreciated not only by me but also by the reader.

A large number of other academics have contributed greatly to this book. My thesis supervisors, Mr Pramit Chaudhuri and Mr Daniel Jones, helped me and encouraged me in my efforts. Dr Jorge Katz, Mr Fernando Navajas and Ingeniero Angel Castano gave me invaluable help during my first and difficult field work in Argentina. Mr Thomas Ljung spent many hours discussing not only technical details of machine tools but also strategic issues in the industry producing computer numerically controlled lathes. Dr Daniel Chudnovsky and Mr Masafumi Nagao, with whom I had the pleasure to work for a few months in the Technology Division of UNCTAD, gave me many useful ideas for further work in a critical period of my thesis work. Professor Charles Cooper, to whom special thanks are given, co-supervised me in my thesis work until he left for the Institute of Social Studies in The Hague. He also read in detail and commented extensively upon a late draft of the thesis. Mr Kurt Hoffman, Dr Raphael

Kaplinsky, Mr Howard Rush and Dr Luc Soete read an early draft of the thesis and provided an inspiring working environment during the early phase of the work. Mr Björn Elsässer and I had many interesting discussions on the diffusion of numerically controlled machine tools.

The actual work on the thesis and this book has been undertaken as part of the research programme entitled 'Technology and Development' at the Research Policy Institute. My colleagues at the institute have provided a stimulating environment. In particular, I would like to thank Dr Charles Edquist, Mr Bo Göransson and Mr Gunnar Paulsson for reading and commenting on earlier drafts.

A book like this is impossible to write without the help of a great number of representatives from firms. For reasons of confidentiality, the names of the firms and their representatives will not be mentioned, but my thanks are still extended to them.

Various branch organizations, in particular CECIMO, VDW and MTTA, gave me a lot of help with regard to basic production and trade statistics. Ms Eva Johansson and Ms Yael Tågerud undertook the typing of this book in a most professional way.

The Swedish Agency for Research Cooperation with Developing Countries (SAREC) financed most of the work going into this book. Their support is gratefully acknowledged.

List of Abbreviations

BTN	Brussels Tariff Nomenclature
CAD	Computer Aided Design
CAFMHA	Camara Argentina de Fabricantes Maquinas Herramienta
CCCN	Customs Cooperation Council Nomenclature
CECIMO	Comité Européen de Coopération des Industries de la Machine-Outil
CEPD	Council for Economic Planning and Development
CETDC	China External Trade Development Council
CNC	Computer Numerical Control
DGFM	Direccion General de Fabricaciones Militares
ECLA	Economic Commission for Latin America
FMM	Flexible Manufacturing Module
FMS	Flexible Manufacturing Systems
FRRF	Fixed Rate Relending Facility
GDP	Gross Domestic Product
IDIC	Industrial Development and Investment Center
INDEC	Instituto Nacional de Estadisticas y Censos
INTI	Instituto Nacional de Tecnologia Industrial
ISIC	International Standard Industrial Classification
ITRI	Industrial Technology Research Institute
JETRO	Japan External Trade Organization
JMTBA	Japan Machine Tool Builders' Association
KAIST	Korea Advanced Institute of Science and Technology
KMTMA	Korea Machine Tool Manufacturers' Association
Korea	The Republic of Korea
LAFTA	Latin American Free Trade Organization
MCI	Ministry of Commerce and Industry
MIRL	Mechanical Industry Research Laboratories
MTTA	Machine Tool Trades Association
NC	Numerical Control
NCMTs	Numerically Controlled Machine Tools
NMTBA	National Machine Tool Builders' Association
NICs	Newly Industrializing Countries
NT	New Taiwanese Dollars
OECD	Organization for Economic Cooperation and Development

R&D	Research and Development
RAM	Random Access Memory
ROM	Read Only Memory
SEK	Swedish krona
SITC	Standard International Trade Classification
Taiwan	The Republic of China
UK	United Kingdom
UN	United Nations
UNCTAD	United Nations Conference on Trade and Development
USA	United States
USD	United States dollar
VDW	Verein Deutscher Werkzeugmaschinenfabriken e.V.

List of Tables

Table		page
2.1	Share of metal products in manufacturing output (%)	6
2.2	Stock of metalcutting machine tools in Japanese, US and UK engineering industries	7
2.3	Average price per unit of CNC lathes produced in Japan, USA and the FRG, 1975–1983 (current US$'000)	10
2.4	Price ratios of units purchased of CNC lathes and conventional lathes in Japan, 1974–1984	11
2.5	Investment in CNC lathes as a percentage of all investment in lathes in a number of OECD countries	13
2.6	Non-socialist world production of lathes	16
2.7	Annual rates of growth of investment in CNC and conventional lathes in a number of OECD countries (in constant prices)	17
2.8	Annual investment in various types of lathes in Japan, 1973–1983 (in millions of 1975 yen and %)	17
2.9	Investment calculation (1) assuming one-shift operation	20
2.10	Investment calculation (1) assuming two shifts for CNC lathes and one shift for conventional lathes	20
2.11	Investment calculation (2) assuming one-shift operation only	21
2.12	Choice between CNC lathes and conventional lathes in Taiwan assuming one-shift operation (US$)	26
2.13	Choice between CNC lathes and conventional lathes in Taiwan assuming two-shift operation (US$)	26
2.14	Break-even wage for the choice of CNC lathes in Taiwan (wage p.a. in US$)	27
2.15	Break-even wage for the choice of CNC lathes in Argentina (wage p.a. in US$)	27
2.16	Choice between CNC lathes and conventional lathes in Argentina assuming one-shift operation (US$)	28
2.17	Choice between CNC lathes and conventional lathes in Argentina assuming two-shift operation (US$)	28
2.18	Illustrative estimate of demand for CNC lathes in the developed market economies and in some NICs	29
3.1	The production of CNC lathes in Japan, Europe and the USA (in US$m. and as % of world production)	32

LIST OF TABLES

3.2	The production of CNC lathes in Japan, Europe and the USA (units)	33
3.3	The Japanese share of the non-Japanese world market for CNC lathes (%)	34
3.4	Production of CNC lathes by individual European countries (in units and as % of total European production)	35
3.5	The principal trade flows in CNC lathes in 1975, 1980 and 1983 (current US$m.)	36
3.6	Investment in NCMTs in large and small enterprises in Japan, 1970–1981 (yen m.)	45
3.7	Weight per CNC lathe produced in Japan and in the FRG (tons)	48
3.8	Market profile for horizontal CNC lathes in the USA with respect to the strength of the motor, 1980	49
3.9	Concentration of production of CNC lathes in Japan, 1970–1981 (in value, units and %)	50
3.10	Sales and profits of five Japanese CNC lathe producers	51
3.11	The distribution of the stock of NCMTs by size group of firms in the FRG, 1980	54
3.12	The size of the domestic market for CNC lathes in the USA, Japan, UK, Italy, France and FRG, 1976, 1980, 1983 and 1984 (in value and units)	59
3.13	Japanese trade in CNC lathes, 1969–1984 (units)	60
3.14	Production and trade in CNC lathes in the FRG and France, 1976	61
3.15	Exports as share of value of production of CNC lathes in four EEC countries, selected years, 1976–1984 (%)	62
3.16	The ratio of external capital to total capital in some CNC lathe producers (%)	64
3.17	Debts as a percentage of total capital in some CNC lathe producers	64
3.18	Share of all imports and those of Japanese origin in the apparent consumption of CNC lathes in some OECD countries, 1980 and 1984 (% of value)	68
3.19	The international specialization and world market share of the Swedish CNC lathe industry, 1974–1984	74
3.20	The international specialization and world market share of the UK CNC lathe industry, 1975–1984	78
3.21	Shipments and unfilled orders of metalcutting machine tools in the USA (US$m.)	80
3.22	Summary of the main characteristics of the three strategies pursued by OECD-based CNC lathe producers	84
3.23	Ranking of the barriers to entry involved in the three	

	different strategies pursued by OECD-based firms producing CNC lathes	85
4.1	Structure of costs for CNC lathes and engine lathes in one NIC-based firm (excluding sales costs and profits) (%)	91
4.2	Price advantages as the annual level of demand for control systems (incl. motor) increases, as suggested by a large supplier	93
4.3	Price advantages as the annual level of demand for CNC units increases, as suggested by a leading European producer of CNC lathes	93
4.4	Estimated maximum price reductions attainable for components (%)	94
4.5	An estimate of the size and origin of economies of scale in CNC lathe production (%)	98
5.1	Production, exports, imports and demand for machine tools in five NICs, 1983	102
5.2	Production of and demand for CNC lathes in five NICs (units)	103
5.3	Summary of the main characteristics of the low-performance strategy	104
5.4	Selected characteristics of eight NIC-based firms producing CNC lathes	106
6.1	Growth in value added in the engineering industry and in the manufacturing sector in Argentina, selected years 1950–1979	114
6.2	The structure of value added in the engineering industries in Argentina, the FRG, Sweden, the UK and the USA (%)	115
6.3	Investment in machinery, equipment and transport equipment and the domestic content of investment in Argentina, 1971–1979	116
6.4	Production and trade in machine tools in Argentina, 1962–1983	117
6.5	Price per unit of machine tools traded in Argentina, 1962–1978	118
6.6	Destination of exports of machine tools from Argentina, 1976	118
6.7	Structure of costs of metalcutting machine tools in Argentina in the early 1970s (%)	119
6.8	Fluctuations in the real exchange rate in Argentina, 1968–1979 (%)	122

6.9	The instalment year of the known CNC lathes in Argentina, 1976–1981 (units)	123
6.10	The '1976 potential' for CNC lathes in Argentina (units)	126
6.11	An estimation of yearly demand for CNC lathes in Argentina, 1981–1985 (units)	127
6.12	Production of lathes in Argentina, 1957–1979 (units)	129
6.13	Price per unit of lathes exported and imported in Argentina, 1975–1978 (US$)	129
6.14	An estimation of advantages from specialization and scale in an Argentinian firm producing CNC lathes	133
6.15	Sources of scale advantages (estimated) with a production of 30 CNC lathes per month in an Argentinian firm	134
6.16	Cost structure (estimated) with a production of 30 CNC lathes per month in an Argentinian firm	134
7.1	Production value in the manufacturing industry and the share of the engineering sector in Taiwan, 1976, 1980 and 1984	144
7.2	The structure of the value of output in the Taiwanese engineering industry, 1976 and 1984 (%)	144
7.3	The share of exports in GDP in Taiwan, selected years 1961–1983	144
7.4	Investment in machinery, equipment and transport equipment and the import content of investment in Taiwan, 1970–1980	145
7.5	Production and trade in the engineering industry in Taiwan, 1976 and 1984	145
7.6	Production and trade in engineering products in Taiwan at sector level, 1976 and 1984	147
7.7	Production and trade in machine tools in Taiwan, 1969–1983	148
7.8	The destination of exports of machine tools and lathes from Taiwan, 1981	149
7.9	Price per unit of lathes traded in Taiwan, 1969–1984 (US$)	150
7.10	Price per unit of horizontal engine and toolroom lathes, non-NC, in US imports, 1980	150
7.11	Nominal tariffs for lathes in Taiwan, selected years 1948–1981	151
7.12	Fluctuations in real exchange rates in Taiwan and Argentina, 1968–1979	152
7.13	Production and trade in CNC lathes in Taiwan, 1977–1984 (in units and value)	153
7.14	Production and trade in lathes in Taiwan, 1969–1984 (units)	155

LIST OF TABLES

7.15	Production and trade in lathes in Taiwan, 1969–1984 (value)	156
7.16	The ratio between consumer goods and engineering goods production in Taiwan, 1965 and 1982 (%)	164
8.1	The share of imports and exports in GNP in Korea, selected years 1960–1978 (%)	174
8.2	The ratio between consumer goods and engineering goods production in Korea, 1960, 1975 and 1981 (%)	175
8.3	Value added in the manufacturing industry and the share of the engineering sector in Korea, selected years 1960–1982	175
8.4	Production and trade in engineering products in Korea, 1980	176
8.5	Production and trade in machine tools in Korea, selected years 1971–1984	178
8.6	Production and trade in lathes in Korea, selected years 1974–1984	178
8.7	Price per unit of lathes traded in Korea, selected years 1974–1984 (US$)	179
8.8	Destination of exports of lathes from Korea, 1977, 1979, 1981 and 1984 (% of value)	179
8.9	Fluctuations in the real exchange rate in Korea, 1967–1978	180
8.10	Production, trade and size of the local market for CNC lathes in Korea, selected years 1974–1984 (in units and value)	183
8.11	Production of machine tools by firm B, 1977–1983 (in units and value)	184
8.12	Value of production by product in firm D, 1 January – 31 July 1985	187
9.1	Production of machine tools in Argentina, Korea and Taiwan 1969–1983, selected years (US$ m.)	202
9.2	Exports of machine tools from Argentina, Korea and Taiwan 1969–1983, selected years (in value and % of production)	203
9.3	Production of lathes in Argentina and Taiwan, 1969–1979 (units)	203
9.4	Production and exports of lathes in Korea and Taiwan, selected years 1974–1984 (US$m.)	204
9.5	Production of CNC lathes in Argentina, Korea and Taiwan (units)	204

xx   LIST OF TABLES

9.6 Calculation of subsidy of leading Korean firm in the CNC
     lathe industry   206
9.7 The main state-controlled variables influencing firm
     behaviour   217

List of Figures

Figure   *page*
2.1 Choice between different types of lathes in the FRG in the
     early 1970s   14
3.1 The S-curve for CNC lathes in Japan, Sweden and UK   43
3.2 Yearly shipment of numerical control units by US
     manufacturers and by Fanuc, 1964–1974   56
3.3 A strategic map of the CNC lathe industry   82
5.1 Map of the position of eight NIC-based firms within the
     low-performance strategy   108
7.1 Sales of engine lathes for firm A, 1973–1980   157
7.2 Sales of NC machine tools for firm A, 1975–1980   158
9.1 The cost, benefit and duration of infancy   214

1

Introduction

1.1 The subject matter and its relevance

This book addresses the question of the adjustment of firms to rapid technical change and the role of government policies in this process. The industry chosen for empirical work is the computer numerically controlled (CNC) lathe industry, and the countries that have been investigated are the Newly Industrialized Countries (NICs) Argentina, Taiwan and the Republic of Korea (henceforth Korea), and Japan as well as some European countries. The book is thus an analysis of the global CNC lathe industry.

The book has its intellectual origin in two separate debates current at the time of formulating the research question, namely at the end of the 1970s and in the early 1980s. The first was the debate about the industrial impact of the 'electronic revolution'. Initially, the concern was about the effects on the OECD countries, mainly in terms of employment effects but also in terms of industrial restructuring (see, e.g., Barron and Curnow, 1978; Freeman, 1977; Jenkins and Sherman, 1979; McLean and Rush, 1978). Slowly, however, the focus was enlarged to include the developing countries (see, e.g., de Arocena, 1981; Hoffman and Rush, 1980; Jacobsson and Sigurdson, 1983; Kaplinsky, 1982 a,b; Lahera and Nochteff, 1983; Rada, 1980; Tauile, 1981).

The second debate focused on the place of the developing countries within the international division of labour in innovation. In particular, how governments can strengthen the position of the developing countries was debated (Cooper, 1976; Cooper and Hoffman, 1978; Katz, 1976; Lall, 1978; Rosenberg, 1976).

The subject matters of the two debates were then integrated into the main research question of the book: under what conditions will the leading lathe producers in Argentina, Taiwan and Korea be able to switch from the production of conventional lathes to computer numerically controlled lathes (henceforth CNC lathes) and is there a role for government to help the firms to adjust? The book is, however, also an analysis of the global industry, and a large part of the book deals with the changing structure of the international CNC lathe industry. Some emphasis, in the developed country part of the book, is given to Japan and to the smaller Western countries – Sweden and the UK. Hence, the book is a case study of the relationship between electronics-related radical technical change and global restructuring of the lathe industry. Some emphasis is given to the role of the government in the adjustment process.

The book is empirical on account of my desire to find out what really happens in the world division of labour in innovation when a product is affected by radical technical change. At the time of the initiation of the work, the literature on the impact of electronics on the developed countries, not to speak of the developing countries, was sparse and highly speculative. I therefore felt there was a need for in-depth studies at the product level. The book can therefore be seen as a contribution to the literature on how the global industry is being affected by the 'electronic revolution'.

The book is, however, also of some theoretical interest. It touches on the question of how an industrial policy should be designed in order to foster, in this case, the production of CNC lathes in the NICs as well as in the UK and Sweden. As Arnold (1983, p. 1) notes: 'Much of our economic thinking depends upon assumptions about firm behaviour.' Explicitly, but normally implicitly, a certain firm behaviour is assumed in various policy prescriptions to governments as regards the development of industrial and technological capabilities. The efficiency of the policies adopted would then, partly, be a function of the validity of the assumptions made concerning firm behaviour. The most interesting result of my work, from a theoretical point of view, is therefore that I have linked the theoretical framework of analysing the nature of competition in an industry with an analysis of how the government should design its intervention policies (Chapter 9).

I would like to emphasize that I made the decision to conduct the analysis with a very strong emphasis on the firm level in order to understand how *actual* firm behaviour was being influenced by state policy. The generation of industrial and technological capabilities within firms is a function of the decisions taken by management on the strategy for the firm. A number of strategies are often theoretically possible for firms to pursue. The pursuit of each strategy involves the development of particular characteristics, e.g. skills, of the firm. In order to understand properly the performance of firms and industries and how this is influenced by the context in which they operate, one therefore needs to understand the management's reasoning as regards the strategic choice. It is only then that one can say something concrete about the efficiency of various types of state policies.

Apart from this general methodological point, the book can also be of interest, in terms of method, from a more practical point of view. When analysing an industry in the degree of detail that is attempted in this book, one obviously needs a set of tools. Such tools may be thought to be the various models of market structure found in standard economic textbooks. Normally, however, the decision variables for the firm or entrepreneur in these models are restricted to the price and quantity of the product of the firm. Firms are assumed to react with only these decision variables. Approaching the real world, i.e. firms producing lathes of various kinds, with these models in mind, resulted in a somewhat confused researcher. A need for a better theoretical framework was obvious. A review of some of the literature on the border between economics and business administration, as well as going back to Chamberlain (1960), resulted in a theoretical framework that proved useful in the subsequent analysis of the nature of competition in the industry producing CNC lathes (see section 3.2). It is to be expected that the framework can be useful in analysing other industries where product differentiation is an important part of the industrial structure.

1.2 Method and structure of the book

The main empirical work in this book was done at the level of a product, the lathe, and at the level of the firm(s) producing this product. In order to be able to pursue the analysis at this level, I have had to acquire a fair knowledge of the engineering aspects of lathes. I had to, for example, learn the main variables used in differentiating a firm's lathes from its competitors'. In order to learn about the technology and the factors influencing its diffusion, I sought information from a number of sources:

- I spent a large number of hours with a mechanical engineer at the Lund Institute of Technology, Sweden;
- I visited four international machine tool exhibitions for several days each: Birmingham, UK; Stockholm, Sweden; Buenos Aires, Argentina; and Seoul, Korea;
- I read a number of engineering journals on a regular basis – in particular, I would like to mention the Japanese journal *Metalworking, Engineering and Marketing* and the US journal *American Machinist*;

- I interviewed about a dozen users of machine tools, mainly in Sweden and Argentina but also in Taiwan.

In Chapter 2, I introduce the technology to the reader in a non-technical way, discuss its evolution over time, analyse the factors influencing its diffusion and present data on its actual diffusion in a number of countries.

In Chapters 3 and 4, I describe and analyse the structure of the international CNC lathe industry. Special emphasis is given to the Japanese and European industries. The central pieces of information on firm behaviour and the nature of competition in the industry were provided by firms that either produce or distribute machine tools. All in all, eleven producers of computer-controlled lathes in Europe and Japan were interviewed. Altogether these firms produce around 8,000 CNC lathes annually. The European firms interviewed accounted for around 60 per cent of total European production in 1984 and the Japanese firms interviewed (in 1983) accounted for 39 per cent of Japanese production in 1981. Among the firms we find the second and third largest producers in the world but also comparatively small firms. A large number of distributors in Europe were also interviewed, as well as firms (producers and distributors) present at machine tool exhibitions. Considerable use was also made in these chapters of machine tool production data supplied by CECIMO and various machine tool trade associations, as well as technical brochures from firms.

Chapter 5 presents a summary of the position of eight CNC lathe producers in the NICs within the global industry producing CNC lathes. All in all, twelve producers of lathes were interviewed in three NICs – Argentina (2), Taiwan (7) and Korea (3). One of the Argentinian firms did not produce CNC lathes and one Taiwan firm did not provide enough data to be analysed properly. Several distributors of machine tools were also interviewed.

In Chapters 6, 7 and 8, case studies of the three countries are presented in some detail. Apart from describing and analysing the firms producing CNC lathes, I analyse the (local) context of the firms. In particular I emphasize the role of government policies and the size of the domestic market for CNC lathes. I attempt to trace the relationships between past government policies and the historical generation of resources within the firms, which have led to their present strategic position within the international CNC lathe industry. I also, however, describe the present government policies and analyse their appropriateness both from the viewpoint of the firms and from the viewpoint of the society. Apart from the sections dealing with the lathe producers, I rely in these chapters to a large extent on secondary sources. Some interviews with government officials and industry associations were however also made.

In Chapter 9, I return to one of the two starting points of the work, namely the role of government policies in strengthening the position of the developing countries within the international division of labour in innovation. I survey some influential literature on the issue of state intervention. Thereafter, on the basis of my theoretical framework for analysing the behaviour of firms, and the case study of CNC lathes, I proceed to draw my own conclusions for government policies. Whilst the bulk of the policy discussion refers to the NICs, I also discuss firm performance and implications for government policy in the European industry, in particular the Swedish and the UK industry.

In Chapter 10, I provide a brief summary of the book.

2

The Technology and its Diffusion

2.1 Introduction

Modern society is to a very considerable extent a machine-based society. A specialized machinery-producing industry was developed in the nineteenth century and the share of the metalworking industry in manufacturing output grew from approximately 10 per cent at the turn of the century to more than 40 per cent today (see Table 2.1). Of particular importance to the development of the machinery industry were innovations in the field of *machines to make machines* or machine tools. The central role of the machine tool industry in enabling the development of a specialized machinery industry was analysed in detail by K. Marx. One innovation of special importance, according to Marx, was that of Henry Maudsley, who produced the first lathe. With this lathe, it was possible for the first time to produce geometrically accurate straight lines, circles, cylinders, cones and spheres required for the parts of machines (Marx, 1974, p. 363). Other authors, e.g. Landes (1969) and Rosenberg (1976), also emphasized the role of the machine tool industry in diffusing new innovations in the nineteenth century.

The central production process in metalworking is that of shaping the metal into useful components with specific dimensions, shapes and finish. There are many metal-shaping methods but the one that probably produces the widest variety of products is the machining process.

Table 2.1 *Share of metal products in manufacturing output (%)*

	1900	1913	1937	1955	1978
USA	10	13	31	41	44
UK	16	19	29	38	41
Japan	2	15	n.a.	24	40

Source: Research Policy Institute (1983), April, p. 4.

Machining can be done by a number of means. Normally, it involves the application of cutting tools to the raw material, e.g. a steel bar or to a casting. The cutting can be done by a number of different types of machine tools, of which the most important are lathes, milling machines, drilling machines, sawing machines and grinding machines.

In a lathe, the cutting tool is applied to the work piece, which is rotated, with the result that metal chips are cut off the work piece. In other machine tools, the cutting tool may be the rotating part, e.g. in a drilling machine.

Among the machine tools used in industry, lathes are perhaps the most important. Data on the stock of metalcutting machine tools in the Japanese, US and UK industries are presented in Table 2.2. Lathes accounted for 19.5–24.6 per cent of the total number of metalcutting machine tools in these three countries. Of course, there are problems with using the number of machine tools as the unit of measurement as machine tools are very heterogeneous with respect to many characteristics, especially their unit value. When possible, I shall use value as the unit of measurement. Value per unit will also be used in some cases when appropriate.

Table 2.2 *Stock of metalcutting machine tools in Japanese, US and UK engineering industries*

	Japan (1981)		USA (1983)		UK (1983)	
	Units	% of stock	Units	% of stock	Units	% of stock
Lathes	139,953	22.2	332,327	19.5	174,374	24.6
Drilling machines	118,811	18.9	281,453	16.5	151,330	21.3
Grinding machines	99,936	15.9	383,027	22.5	113,216	16.0
Milling and planing machines	69,576	11.1	218,479	12.8	85,460	12.1
Special machines	68,649	10.9	n.a.	n.a.	n.a.	n.a.
NC machines	22,397	3.6	92,772	5.4	25,802	3.6
Others	109,366	17.4	394,775	23.2	158,245	22.3
Total	628,688	100.0	1,702,833	99.9	708,427	99.9

Sources: Digest of Japanese Industry and Technology (1982); *American Machinist* (November 1983); MTTA (1985).

In section 2.2, I shall explain in more detail the technology of CNC lathes and how it evolved in the 1970s. In section 2.3, I shall analyse the choice of technique in turning or the choice between using different types of lathes. This section includes not only a discussion on which type of lathe will be chosen in different contexts but also some evidence on how that choice has altered over the last ten years. The factor-saving bias of CNC lathes is analysed in section 2.4 in relation to conventional lathes. Finally, in section 2.5, I shall draw some conclusions for lathe producers in the NICs concerning their entry into production of CNC lathes. The discussion will be based on a market estimate for different types of lathes.

2.2 CNC lathe technology

Production in the engineering sector may be divided into two main areas characterized by different technical conditions of production. The production of long series of standardized products, e.g. engines, has for some decades been subject to mechanization efforts. Large production volume justifies extensive development of expensive, purpose-built and inflexible equipment, e.g. transfer lines.

The second area is characterized by batch production. For products with relatively low levels of standardization, the demand for flexibility is very high, and massive investments in special-purpose, dedicated production systems do not make economic sense. Instead, batch production was undertaken by a set of standard, multipurpose and hand-operated, machine tools. This ensured a high degree of flexibility but, at the same time, the advantages of automated systems were not reaped.

In the operation of a machine tool one may identify the following sequence:

(1) the work piece is transported to the machine
(2) the work piece is fed into the machine and fastened
(3) the right tool is selected and inserted into the machine
(4) the machine is set (for e.g. operation speed)
(5) the movement of the tool is controlled
(6) the tool is changed
(7) the work piece is taken out of the machine
(8) the work piece is transported to another machine tool or to be assembled
(9) the whole process is supervised in the case of, e.g., tool failures.

Traditionally, all these elements in batch production were performed manually. As a result, batch production was associated with very high costs; high and even quality was difficult to achieve; it was often tedious to set the machine and change the tools, with the consequence that the level of utilization of the machine tool was very low. Furthermore, very skilled people were needed to perform these functions.

At the beginning of the 1950s, the first numerically controlled machine tool (NCMT) was developed at the Massachusetts Institute of Technology with the financial backing of the United States Air Force. Instead of having a worker performing tasks 4 (some of the setting tasks still need to be done manually) and 5 above, the information needed to produce the part was put on a medium, for example a tape, and fed into the numerical control unit. By simply changing the control tape, the numerically controlled machine tool could quickly be switched to the next job, which could involve a totally different sequence of operations. This meant that flexibility and automation could be combined.

However, the technology failed to be diffused on a wide scale for several reasons. First, the logic in the control units of these first numerically controlled machine tools was made of hardwired circuitry and, if new functions were to be performed, a change in the hardware had to be made. Secondly, the components were very unreliable and, thirdly, very costly. It was not until around 1970, when these relatively inflexible numerical control units were replaced by *softwired* mini computers, that the diffusion of NCMTs took place on a larger scale.

A still more significant change in the technology, which indeed could be looked upon as a great qualitative change as opposed to a change in degree, was the use of microcomputers for the numerical control unit. This technology began to be incorporated in the NCMTs in 1975–76, only a few years after the development of the micro *chip*. Indeed, Dr Inaba, President of the very successful Japanese firm Fanuc, which pioneered this development, stated: 'We applied the technical innovations in the semiconductor field to machine tools earlier than the computer industry did' (*Metalworking, Engineering and Marketing*, 1979, p. 42).

In contrast to the hardwired numerical control units of the 1960s, these computer numerically controlled (CNC) units have a memory that contains the control programme. Two types of CNC units have been developed (*Metalworking Production*, 1979):

- *Fully flexible*, which are practically the same as mini computer based CNC units but constructed with Large Scale Integrated Circuit (LSI) chips. The control programme is contained in a random access memory (RAM) and it can be updated or modified at any time.
- *The fixed feature* system is equivalent to the *hardwired* numerical control units in that the system performance and features are designed in software in a read only memory (ROM), which cannot be changed except by changing the chip. The software capabilities can therefore be said to be more limited than in case of the *fully*

flexible CNC units. This type of CNC unit is designed to meet the specifications of the vast majority of customers. Typically, these are two-axis systems for lathes and three-axis systems for milling and drilling machines. It is this second type of control unit that has had a tremendous effect on the diffusion of numerically controlled machine tools and, above all, CNC lathes.

The shift to computer-based, and particularly to microcomputer-based, NC units has had several important implications for the NCMTs:

(1) The versatility and flexibility of the machine tool has been considerably enhanced owing to the computer's ability to store and edit programmes.
(2) The price of the NC unit has been reduced from roughly 30–35 per cent to 15–20 per cent of total costs. The price reduction has been particularly important for CNC lathes, which are relatively small machine tools and therefore cheap in comparison with, say, a milling or a boring machine. Prior to the introduction of microcomputer-based CNC units, the control unit constituted a very sizeable share of the total cost of a large CNC lathe. Indeed, the availability of cheap CNC units has permitted the development of smaller CNC lathes too, a fact of great relevance to the process of diffusion of CNC lathes in the past decade. The Japanese producers have been particularly successful in developing smaller CNC lathes. This is strikingly illustrated in Table 2.3, where the average prices per unit of Japanese, US and

Table 2.3 *Average price per unit of CNC lathes produced in Japan, USA and the FRG, 1975–1983 (current US$'000)*

	Japan	USA	FRG
1975	48.3	129.6	150.7
1976	43.3	155.8	192.7
1977	48.3	165.7	145.9
1978	55.1	162.0	126.3
1979	54.6	147.5	181.0
1980	55.9	175.5	171.6
1981	60.3	218.4	136.0
1982	54.4	223.8	148.7
1983	51.0	188.4	163.0

Source: Elaboration of data supplied by the Comité Européen de Coopération des Industries de la Machine-Outil (CECIMO), NMTBA (1984/5), and the Verein Deutscher Werkzeugmaschinenfabriken e.V. (VDW).

West German CNC lathes are shown. The price of the Japanese CNC lathes is one-third of the price of the CNC lathes produced in the United States and in the Federal Republic of Germany (FRG). The price differences reflect mainly the differences in the size of the CNC lathes, the Japanese ones being much lighter and smaller. As will be discussed further in Chapter 3, the large output of smaller Japanese CNC lathes has greatly contributed to the present diffusion of CNC lathes to small and medium-sized firms, which now have an *on the shelf* technology that corresponds to their requirements.

The reduction in the price of the control unit has also contributed to the improved competitive position of CNC lathes vis-à-vis conventional lathes. As is indicated in Table 2.4, the price ratio of units purchased in Japan of CNC lathes and conventional lathes dropped from 8.3 in 1974 to 2.9 in 1981. Since 1981, the ratio has increased again slightly to 3.5. Similarly, for the *production* of lathes in the United States, Japan, the Federal Republic of Germany, France, the United Kingdom and Italy, the ratio of price per unit of CNC to conventional lathes dropped from 7.3 in 1976 to 3.2 in 1981. Here, too, the ratio rose somewhat thereafter, to approximately 3.6 in 1984.

Table 2.4 *Price ratios of units purchased of CNC lathes and conventional lathes in Japan, 1974–1984*

	Price per unit		Price ratio of CNC to conventional lathes
	Conventional lathes (yen m.)	CNC lathes (yen m.)	
1974	2.07	17.20	8.31
1975	2.98	14.46	4.85
1976	2.43	11.75	4.84
1977	2.25	9.82	4.36
1978	2.59	11.10	4.29
1979	2.25	9.55	4.24
1980	2.75	9.57	3.48
1981[a]	3.08	8.93	2.90
1982	3.29	10.63	3.23
1983	2.77	9.80	3.54
1984	3.29	11.65	3.54

Sources: 1974–80: elaboration of data supplied by Japan External Trade Organization (JETRO); 1981: elaboration of data supplied by CECIMO and the Japan Machine Tool Builders' Association (JMTBA); 1982 and 1983: elaboration of NMBTA (1983/84, 1984/85).

[a] Assuming no imports of CNC lathes. See Table 3.13.

(3) The programming has been simplified to the point where, for the least complex CNC lathes, practically anybody can do the programming with only a minimum of training and with no help from expensive programming equipment. In fact, two Japanese firms have developed a control unit that asks the operator for instructions and the operator only needs to press a button to answer (*Japan Economic Journal*, 9/6 1981, 23/6 1981 and interview with Japanese producer of CNC units). Other firms have now also developed such units.

(4) Given that a computer is attached to the machine tool, the use of electronics to control functions other than tasks 4 and 5 (setting and tool control) have been made possible. Thus, it is now normal for tool changing (tasks 3 and 6) to be done automatically. Furthermore, robots are being attached to or integrated in the CNC lathes, performing tasks 2 and 7. Indeed, once a computer controls the machine tool, the performance of several machine tools can be integrated. The integration is beginning to be extended to the design and production planning tasks too. The most important manifestation of these system developments is the flexible manufacturing module (FMM). An FMM is a CNC lathe linked to some kind of robot which does the material handling. Normally, supervisory and other sophisticated control functions are included. An indication of the rate of diffusion of these modules is that around one-quarter of all CNC lathes sold in Sweden in twelve months in 1984/85 were equipped with some kind of robot (Edquist and Jacobsson, 1986). Interviews with leading CNC lathe producers in the UK and the FRG suggest however that the share is much lower in those countries.

While these *systems* developments are the most important features of the developments today, the reductions in the relative price of CNC lathes, their improved reliability and the simplification of the programming have been the main features of the development in the period after 1975. These technical and economic changes have greatly contributed to a very rapid diffusion of CNC lathes in a number of OECD countries, as indicated in Table 2.5.

2.3 Choice of technique in turning

The choice of technique in turning is a complex issue, because not only are there several different types of lathes to choose between, but the products produced by a lathe can vary tremendously in both shape and size as well as in the number of units that are demanded.

Table 2.5 *Investment in CNC lathes as a percentage of all investment in lathes in a number of OECD countries*

Year	France	FRG	Italy	Japan	Sweden	UK	USA
1974	n.a.	n.a.	n.a.	22	34	n.a.	n.a.
1975	n.a.	17	n.a.	23	43	n.a.	n.a.
1976	26	n.a.	15	28	42	19	n.a.
1977	47	n.a.	n.a.	43	53	21	n.a.
1978	n.a.	n.a.	n.a.	41	70	31	n.a.
1979	n.a.	n.a.	n.a.	52	70	38	n.a.
1980	52	47	50	49	69	47	57
1981	n.a.	n.a.	n.a.	45	78	73	n.a.
1982	68[a]	42	58	58	77	79	60
1983	78[a]	n.a.	n.a.	69	72	66	69
1984	85	59	n.a.	71	83	83	73

Sources: Elaboration of data supplied by CECIMO, the various national machine tool builders' associations, and NMTBA (1984/85, 1985/86).

[a] Export of conventional lathes exceeds production. Consumption of conventional lathes is therefore equal to imports.

A large number of factors need to be taken into account by the investor when considering what type of lathe to use. The most important factors are:

- the capital cost of the different types of lathes;
- the cost of labour;
- the cost of cutting the metal, which is a function of the cycle time (the time it takes for the work piece to be machined, excluding the time of preparation), the capital cost of the machine tool, and the cost of the labour time used to operate the machine tool;
- the cost of preparation of the lathes, such as setting and programming;
- the cost of work in progress and inventories.

Abstracting for the moment from the last point, and with given prices for labour and the various types of lathes, the choice of which lathe to use becomes a question of two factors only – the cost of preparation and the cost of the cycle time (the actual cutting cost). The former cost is called fixed, while the latter cost is called variable.

A number of different types of lathes exist which involve different combinations of preparation time cost and cycle time costs in their use.

The *engine* lathe is a simple lathe that is fully manually operated. It is very cheap in comparison with other lathes although the cycle time is long, which implies that the variable costs are high, especially when labour is dear. The set-up time is fairly short though, which implies low fixed costs in its use. Given the characteristics of relatively low fixed costs and relatively high variable costs, the engine lathe is normally used in the production of a small number of small batches. It is found in repair shops as well as in most machine shops as a back-up machine. Figure 2.1 shows that in the FRG in the early 1970s the engine lathe was chosen in areas of operation to the south-west part of the quadrant.

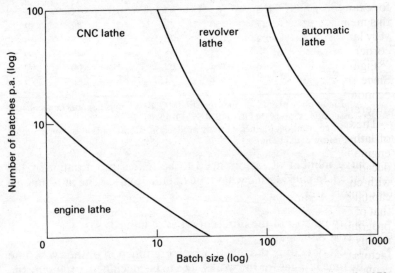

Figure 2.1: *Choice between different types of lathes in the FRG in the early 1970s*
Source: Ljung (1980)

The *CNC lathe* is more complex but much less labour intensive in its operation than the engine lathe. It is also considerably more expensive, even though the gap has narrowed dramatically in the last few years. The cycle time is shorter than for the engine lathe, implying the possibility of lower variable costs, in particular when labour is dear. The fixed, setting and programming costs have been high though, implying higher fixed costs in relation to the engine lathe. The fixed costs of a CNC lathe are, however, somewhat different from those of other lathes as they consist mainly of tape preparation. Hence, it is a *once for all* cost and the cost of preparing the second batch of the same component is very close to nil. This means that CNC lathes are advantageously used instead of engine lathes not only for some larger-sized batches but also when a particular batch is frequently repeated (see Figure 2.1).

The *revolver lathe* is designed to achieve short machining times. Hence, the cycle time is shorter than for the CNC lathe. The revolver lathe is cheaper than the CNC lathe and, with some degree of automation, the variable costs of labour can be kept low. In comparison with CNC lathes the revolver lathe has higher fixed costs but lower variable costs. It is therefore used for larger batch sizes.

Finally, the *automatic lathe* is characterized by very short cycle times but very high fixed costs in setting. An automatic lathe is cheaper than, for example, a CNC lathe and is very labour saving in its operation. All this means that the traditional area for automatic lathes has been the very large batch sizes. Automatic lathes can be found in the northeast corner of Figure 2.1.

Figure 2.1 is based on cost data from the FRG in the early 1970s. Since then the CNC lathes have gone through many changes, both economic and technical, so that the choice of technique would be different today.

The CNC lathe has improved its position vis-à-vis all the three types of lathes shown in the figure. If we compare a CNC lathe with an engine lathe, two things have happened to the CNC lathes in the last decade. First, the CNC lathe has become relatively cheaper compared with conventional lathes, including engine lathes, implying that the capital cost component in the variable cost has decreased relative to that of the engine lathe. Secondly, the preparation time, i.e. the fixed programming time, has decreased dramatically for CNC lathes. This means that the fixed costs have declined drastically too. Both these factors reduce the minimum batch size needed for the CNC lathe to be economic and imply that the CNC lathe has encroached on the engine lathe and extended its frontier to the southwest in Figure 2.1.

Whilst the main advantage of engine lathes vis-à-vis CNC lathes has been their lower fixed costs, the main advantage of revolver and automatic lathes has been their lower variable costs (the costs of the actual cutting time). This lower cost has been a function of both a shorter cycle and lower machine costs. Labour costs for the operation are much the same, as all machines are automatic. The tremendous reduction in the relative price of CNC lathes has meant that the capital cost element in the variable costs has narrowed between the two technologies. This means that the superior flexibility of the CNC lathes, which has been enhanced, provides an increasing competitive threat to revolver and automatic lathes. Thus, both the narrowing variable costs and the widening of the fixed cost differential lead to an extension of the field of application for CNC lathes to the northeast in Figure 2.1.

The change in the competitive position of the different types of lathes is also revealed in the actual investment patterns for these machine tools. In Table 2.6 an approximate indication of the relative

Table 2.6 *Non-socialist world[a] production of lathes*

Year	Conventional (US$ m.)	CNC (US$ m.)	CNC as percentage of total
1975	1,147	445	27.9
1976	1,057	498	32.0
1977	1,132	626	35.6
1978	n.a.	938	35.6
1979	1,515	1,310	46.4
1980	1,625	1,906	54.0
1981	1,554	1,639	51.3
1982	885	1,416	61.5
1983	634	1,280	66.9
1984	558[b]	1,510	73.0

Source: Elaboration of data supplied by CECIMO.
[a] USA, Japan, France, Italy, FRG, UK and Sweden.
[b] Excluding Sweden.

importance of CNC and conventional lathes can be found in the form of the evolution of production of lathes in seven developed countries. These countries together account for more than 80 per cent of the non-socialist world's production of machine tools. It may be seen that the share of CNC lathes in *world* production of lathes grew from about 28 per cent in 1975 to 73 per cent in 1984. The growth in the importance of CNC lathes was also shown in the investment in lathes in a number of developed countries as revealed in Table 2.5. In the USA and Japan more than 70 per cent of the investment in lathes was made in CNC lathes in 1984. In Sweden and the UK, the share was 83 per cent in 1984.

Associated with the rapid substitution of CNC lathes for conventional lathes, the market for the latter has shrunk in both relative and absolute terms. This change in market share is shown in Table 2.7, which indicates that in only one of six developed countries, namely the Federal Republic of Germany, did investment in conventional lathes grow in the second half of the 1970s, in absolute although not in relative terms. This trend continued in the 1980s. For example, in the case of Japan, the demand for CNC lathes in 1984 was 24.3 per cent higher than in 1980 whilst the demand for conventional lathes fell by 48 per cent in the same period.

In terms of the effect on the market for different types of conventional lathes, the case of Japan (see Table 2.8) reveals that the share of engine lathes in the total investment in lathes went down from 31 per cent in 1973 to 12 per cent in 1983 whilst that of other lathes, e.g. revol-

Table 2.7 *Annual rates of growth of investment[a] in CNC and conventional lathes in a number of OECD countries (in constant prices)*

	CNC	Conventional
Japan, 1973–1980	+ 10	− 7
Federal Republic of Germany, 1975–1980	+ 42	+ 5
United Kingdom, 1976–1980	+ 30	− 12
Italy, 1976–1979	+ 52	− 1
France, 1976–1980[b]	+ 30	− 5
Sweden, 1974–1980[b]	+ 12	− 23

Source: Elaboration of data supplied by the various national machine tool builders' associations.
[a] Investment is approximated by production minus export plus import.
[b] Current prices.

ver lathes, decreased from 12 per cent in 1973 to 4 per cent in 1983. The share of automatic lathes was constant until the early 1980s, indicating that CNC lathes had not made any major inroads into the northeast of Figure 2.1 in the 1970s. The situation changed in 1982, however, when the share of automatic lathes fell dramatically.

The very high share of CNC lathes in total investment in lathes in Sweden and in the UK in 1981 and 1982 suggests that this substitution

Table 2.8 *Annual investment[a] in various types of lathes in Japan, 1973–1983 (in millions of 1975 yen and %)*

Year	CNC lathe		Automatic lathe		Engine lathe		Other lathes		Total value
	Value	%	Value	%	Value	%	Value	%	
1973	26,097	22.7	38,583	33.6	36,081	31.4	13,978	12.3	114,738
1974	25,324	24.1	35,251	33.5	28,153	26.8	16,314	15.6	105,042
1975	13,004	23.2	14,623	26.1	21,134	37.7	7,255	13.0	56,016
1976	14,455	29.3	19,494	39.6	10,991	22.3	4,247	8.7	49,187
1977	22,085	42.9	18,533	36.0	7,785	15.1	3,048	6.0	51,451
1978	21,132	41.9	17,250	34.2	8,887	17.6	3,150	6.3	50,419
1979	38,239	51.8	20,711	28.0	12,810	17.3	2,068	2.9	73,828
1980	50,227	48.5	30,959	29.9	15,804	15.3	6,522	6.3	103,512
1981	n.a.	45.0	n.a.	n.a.	n.a.	n.a.	n.a.	n.a.	n.a.
1982	63,084	59.3	22,339	21.0	15,665	14.7	5,180	4.9	106,268
1983	62,486	69.2	13,918	15.4	10,545	11.7	3,372	3.7	90,321

Sources: Based on production and trade data supplied by JETRO and NMTBA (1983/84, 1984/85).
[a] Investment is approximated by apparent consumption, i.e. production minus exports plus imports.

process started earlier in these countries. Indeed, the share of automatic lathes in total investment in lathes in Sweden was already less than 12 per cent in 1981, while the share of engine and other lathes was around 10 per cent. Of particular importance to this process of substitution of CNC for automatic lathes is the trend to attach material handling units or robots to the CNC lathes, which still further reduce the labour content of the CNC lathe and therefore its variable costs. The relative price changes referred to above are also of importance, as well as the increased need for producing shorter series among, e.g., automobile producers in response to customer demand and fiercer international competition.

The fact that CNC lathes have mainly substituted for engine and other, e.g. revolver, lathes is serious for the lathe producers in the developing countries as these have, on the whole, specialized in producing engine lathes. These are the most simple types of lathe to produce. Taiwan, for example, with an export ratio of 78 per cent in 1981, exported 7,958 engine lathes, 446 revolver lathes and 4 automatic lathes (Metal Industrial Development Centre, 1982). Similarly, in 1982, Korea exported 1,558 engine lathes, 17 turret or revolver lathes and 21 automatic lathes (Office of Customs Administration, 1982). It is also in the particular submarket of engine lathes that the developing countries have gained a significant share of a developed country market, namely that of the US market where their share was about 18 per cent in 1980 (UNCTAD, 1982a). This figure has risen since then and in 1983 the share of Taiwan, Korea and India only was around 21 per cent (calculated on the basis of NMTBA-supplied trade data and Office of Customs Administration, 1983).

2.4 The factor-saving bias of CNC lathes

In the previous section I analysed how economic and technical changes altered the choice between the various types of lathes. In this section I shall discuss in detail the effect of choosing CNC lathes, instead of conventional lathes, on the use of various factors of production – capital, skilled labour and semi-skilled labour.

2.4.1 Some examples from Sweden
In order to illustrate how the choice of technique varies between different firms, I shall present some calculations based upon firm interviews in Sweden made in 1981.

First, however, let me introduce the so-called p-factor. The p-factor is frequently used by users of machine tools.

$$p = \frac{\text{cycle time} + \dfrac{\text{preparation time}}{\text{batch size}} \text{ Conventional}}{\text{cycle time} + \dfrac{\text{preparation time}}{\text{batch size}} \text{ CNC}}$$

The p-factor shows the relationship between the use of a conventional and a CNC lathe in terms of the total time per unit of output, both production and setting time. Normally, managers choose the technology that provides the lowest unit cost. The relationship between the p-factor and unit costs is somewhat complex. Let us assume that we are to assess the cost per unit of output when using conventional lathes as opposed to a CNC lathe. The p-factor affects both the cost of capital per unit of output and the cost of labour per unit of output. As can be seen in Table 2.16 below (p. 28), the higher the p-factor, the smaller is the difference between the two technologies in terms of the unit capital cost. This is simply so because, with a higher p-factor, more conventional lathes are required to produce the same output as a CNC lathe than in the case of a lower p-factor. As for labour, to the extent that each conventional machine is tended by one operator, this being the normal case in the firms interviewed, the higher the p-factor, the larger are the savings in labour per unit of output if a CNC lathe is used. Thus, given the factor costs, the higher the p-factor the greater the probability of using CNC lathes.

The first calculation is taken from a firm that produces standardized pumps. Based upon cost data from the firm, I have calculated two examples. The first example shows the case of single-shift production by both CNC and conventional lathes (see Table 2.9). The manager of the firm took only investment costs and labour costs into account in his calculation. He argued that costs for work in progress and for stocks were marginal. The absence of any importance given to these items could be accounted for by the low unit cost of the pumps. On the other hand, he included savings in building costs from using CNC lathes instead of conventional lathes (he calculated a saving of 30 square metres per machine). The reason for saving space is simply that three conventional machines take up more space than one CNC lathe. Given the assumptions of a depreciation time of ten years, an interest rate of 15 per cent and a p-factor of 3, production costs per unit of output with the CNC lathes were only 60 per cent compared with conventional machine tools. Using the CNC lathes implied an increase in capital costs of 27 per cent, and a large decrease in labour costs of 62 per cent.

If, however, we assume that the CNC lathes are used in two shifts and the conventional lathes in one shift, the results become altered

Table 2.9 *Investment calculation (1) assuming one-shift operation*[a]

Cost item	CNC lathes	Conventional lathes
Investment costs p.a.	875,000	560,000
Building costs		+ 126,000[b]
Labour costs	793,968	2,101,680
Total costs	1,668,968	2,787,680

Source: Elaboration of data supplied by the firm.

[a] Other assumptions are: (1) One CNC lathe operator's wage cost is 96,744 SEK; (2) A setter's wage cost is 116,760 SEK. One setter is used for seven CNC lathes; (3) One conventional lathe operator's wage cost is 100,080 SEK; (4) The p-factor is 3; (5) 10-year depreciation and 15 per cent interest rate.

[b] We only know the *extra* cost of buildings associated with the choice of conventional lathes.

somewhat (see Table 2.10). This situation is very common in a country like Sweden where it is normal for operators of conventional machine tools to work only one shift whilst CNC machine tools are operated in two or even three shifts. This difference is a function of the skill content of the operators and hence the relative strength of the unions vis-à-vis the employers. Total costs per unit of output, with CNC lathes in this example are only 52 per cent in comparison with conventional lathes. We can also note that the use of CNC lathes in fact implies a reduction in investment costs per unit of output by nearly 16 per cent while labour costs are reduced by 58 per cent. Hence, socioeconomic factors are important in the choice of technology. We would of course expect these to vary from society to society and it is therefore not possible to generalize about the factor-saving bias of CNC lathes.

Table 2.10 *Investment calculation (1) assuming two shifts for CNC lathes and one shift for conventional lathes*

Cost item	CNC lathes	Conventional lathes
Investment costs[a] p.a.	1,041,000	1,235,000
Labour costs	1,587,936	3,783,580
Total costs	2,628,936	5,018,580

Source: As for Table 2.9.

[a] Assumptions as in Table 2.9, except for the depreciation time which is 7½ years for CNC lathes. I also assumed that the output of the second shift for CNC lathes is only 80 per cent of that of the first shift.

The second firm produces very complex and expensive custom-designed components for the shipbuilding industry. I have calculated only one example, assuming one shift for both type of technologies. It should also be added that the firm had implemented a new work organization simultaneously with the acquisition of the CNC lathes. Instead of having a functional lay-out, which involves putting all machines performing a certain function (e.g. milling) together, they applied the principles of group technology, which involves organizing groups of machine tools that together can perform all the required functions on a work piece and for which the use of CNC lathes is all but a prerequisite.

The investment data can be found in Table 2.11, from which it may be seen that, in this case, work in progress and stocks account for a large part of the capital cost. This is explained by the expensive components that the firm includes in its final product. With group technology, the throughput of products can be faster than in a functional lay-out. The faster throughput also means that the stocks of components can be reduced. Total costs are thereby decreased by 15 per cent. The reductions in the cost for stocks and work in progress meant in this case that, even with one-shift operation, the use of CNC lathes was slightly capital saving, by 4 per cent. The savings in labour amounted to 41 per cent. The lower reduction in labour costs per unit of output compared to Tables 2.9 and 2.10 is accounted for by the fact that a p-factor of 2 was claimed to operate in this instance, whilst it was 3 in the prior case.

Table 2.11 *Investment calculation[a] (2) assuming one-shift operation only*

Cost item	CNC lathes	Conventional lathes
Investment costs p.a.	2,275,000	1,263,000
Cost for work in progress and stocks	2,400,000	3,600,000
Labour costs	1,296,870	2,201,760
Total costs	5,971,870	7,064,760

Source: Elaboration of data supplied by the firm.
[a] Assumptions as in Table 2.9, except for a p-factor of 2. Space savings were not calculated by the firm.

2.4.2 *The effects on the use of the various factors of production*

(i) Investment costs It appears as if the investment costs per unit of output may still sometimes be higher for CNC lathes than for conventional lathes. It may be mentioned, however, that in another firm visited it was claimed that the investment costs were the same even though they operated with only one shift. The firm representative also claimed that the costs for work in progress and stocks were reduced by 50 per cent. It should also be mentioned that, in Table 2.10, the investment analysis was made on the assumption that the cheapest East European conventional lathes were bought.

The flexibility of CNC lathes in relation to, for example, revolver lathes means that stocks can be reduced (through reducing the batch sizes). Hence, firms can keep smaller stocks of intermediate and final products since the flexible production system can quickly respond to new demands. The faster throughput of the products in a CNC lathe, especially in relation to engine lathes, also contributes to a reduction in the costs for work in progress. This is of course, as was shown in the above examples, of varying importance depending on the cost of the components produced, i.e. the value of the work in progress. The reductions in the cost of stocks and work in progress can however be of some importance: in the case of the engineering industry in Sweden in 1977, 31.5 billion SEK were tied up in goods while only 15–20 billion SEK were tied up in machines and buildings (DEK, 1981).

The example in Table 2.10 also indicated that the considerable space savings that may be achieved when CNC lathes are used can have some influence on the investment cost even though a forty-year depreciation time was used for the buildings. Savings in space can also be of importance for other reasons. In Buenos Aires in Argentina, for example, expansion of factory space is prohibited by law. Hence, if a firm wants to expand its output, automated machine tools are the only solution if it does not want to move away from the city and its workforce.

It is difficult to give a general answer to the question of whether CNC lathes are capital using or capital saving. It seems as if it is partly a matter of the size of reductions in the cost for work in progress and for stocks. It is also a function of the size of the p-factor, which is partly a function of the type of product produced. With a higher p-factor, the higher are the investment costs for conventional lathes in relation to CNC lathes. A further factor to take into consideration, which indeed appears to be decisive in a country like Sweden, is the prevailing institutional restriction on the full utilization of capital in the form of a lack of skilled workers willing to work shifts with conventional machine tools. In the Swedish context, CNC lathes can clearly be capital saving and this is to a large extent due to this factor. Indeed, in several of the firms visited the main reason for choosing CNC lathes was said to be the possibility of finding workers who were willing to work shift work with CNC lathes. That CNC machine tools are frequently capital saving in the Swedish context is corroborated by Boon (1984). Six out of eight firms interviewed claimed that machinery investment per unit of output is less for CNC machine tools than for conventional machine tools. In addition, practically speaking all firms claimed that there was a reduction in stocks, work in progress and floor space if CNC machine tools were used.

All in all, we can suggest that CNC lathes can be found to be slightly capital saving in an economy like Sweden, in some cases even if single shifts are assumed for both types of technologies. The factor that ensures that CNC lathes turn out to be capital saving is, however, the overcoming of institutional restrictions on the full use of capital, including buildings, etc., by using CNC lathes. As the institutional restrictions are probably not very strong in the NICs, CNC lathes are probably less likely to be capital saving in these economies than in the developed countries. A further reason for suggesting this is that on the whole I would suggest that the complexity of the products produced with CNC lathes in an NIC, and therefore their value, is less than in a developed country. This would tend to reduce the importance of reductions in work in progress in an NIC.

(ii) **Labour** The biggest source of productivity increase achieved by using CNC lathes lies in the saving of labour. In the examples in Tables 2.9, 2.10 and 2.11, the savings amounted to 40–60 per cent. The degree of labour saving can be even higher as, when the cycle time is long, one operator can serve two CNC lathes. This was indeed the case in one Taiwanese firm. Furthermore, as was noted above, a recent trend is to attach automatic material handling devices to the CNC lathes which reduce the labour requirement still further.

(iii) **Skills** There has been a significant reduction in the skills needed to operate the lathe as well as in the number of people required to embody these skills. An operator of a CNC lathe clearly needs some skills but the maximum time for an unskilled person, with a technically oriented secondary education, to become proficient with a CNC lathe was said by one firm to be 12 months at work, including inhouse courses. Other firms suggested a maximum of six months. By contrast, five years of experience is often mentioned as being necessary to acquire proficiency as an operator of a conventional lathe. In the case shown in Table 2.11, the firm needed to employ only twenty-two semi-skilled CNC lathe operators instead of forty-four skilled operators. In another case, twenty-one CNC lathe operators substituted for sixty-three qualified operators. Hence, the total *mass* of skills needed for the operation of the lathes, has been reduced in a very significant manner.

The use of CNC lathes, however, requires a set of other skills that are not important in the case of conventional lathes. First, in addition to the CNC lathe operators, setters and programmers are needed. About six–eight CNC lathes are served by one setter and one programmer. These people are often former skilled workers who have joined the ranks of the white-collar workers, or technicians.

Second, the repair and maintenance task has become more complex, mainly as regards the electronic part and the interface with the mechanical parts. Given that for most users, apart from very large firms, the electronic part is of a black box character, it is normal for the supplier to take over the repair and maintenance work. Usually the supplier of the numerical control unit provides the service on that unit while the machine tool supplier is responsible for the rest. It is clear that, at the level of the lathe, computerization has increased the skill requirements for repair and maintenance. However, as the number of machine tools is lower in the case of the use of CNC lathes, the *amount* of repair and maintenance skills needed to produce a given number of engineering products has not necessarily increased (Senker *et al.*, 1980).

If we translate the skill requirements into training years, we can calculate the *degree* of skill saving with the use of CNC lathes. With a p-factor of 3 and using the example in Table 2.9, the total number of training years become fifteen for the operation, setting and programming of the CNC lathes and eighty-four in the case of the conventional lathes. Hence the saving in skills amounts to 82 per cent ($(84-15)/84$).

Of course, there is nothing inherent in the CNC technology that dictates the use of semi-skilled labour instead of traditional skilled labour. A skilled labourer can also operate a CNC lathe and will in some cases probably ensure a higher rate of utilization of the CNC lathe than a semi-skilled worker. However, the skill requirements can be reduced and the evidence suggests that on the whole the normal pattern is that semi-skilled workers are used (interviews; Elsässer and Lindvall, 1984; Rempp, 1982).

All in all, CNC lathes can be tremendously skill saving. Proportionately speaking, the savings in skills is greater than that in *undifferentiated* labour, which in turn is far greater than that in capital. This means that the CNC lathe represents a technical change, the appropriateness of which is a function of the degree of scarcity of skilled labour compared to semi-skilled labour and to capital. If we assume that the scarcity of skilled labour in relation to semi-skilled labour is greater in the NICs than in the developed countries, we would be forced to the conclusion that we are witnessing a technical change that has its origin in the developed countries but that is of a greater potential benefit to the NICs. To ascertain if a phenomenon of this kind really is taking place is, however, impossible without detailed data on the relative scarcity of the various factors of production in a range of economies.

In all countries where interviews have been carried out it is claimed that skilled workers are scarce, this includes old industrial nations such as the UK and Sweden. In Argentina, some firms complained that it was difficult to find skilled workers, even in a recession. Other firms claimed, however, that they could be found although when the economy picked up again there would be a shortage since, in the meantime, the skilled workers had left the occupation for other more lucrative work. Hence, the setting of wages appears to be of importance in a country like Argentina, even for the short-term availability of skilled workers. In Taiwan, out of six firms, three claimed it was *difficult* to find skilled operators; one experienced *some scarcity* and two firms said they did not have such a problem.

Balassa (1981) suggests that Taiwan and Korea have an emerging comparative advantage (based on cheap skilled labour) in skill-intensive products, e.g. in some machinery products. To the extent that this is true, a technical change of the kind that CNC lathes represent would keep the developed countries from losing a comparative advantage in these skill-intensive product lines. The argument as to which types of economies would benefit most from CNC lathes would then be reversed.

While admitting the fact that it is an issue that cannot be resolved within the context of this work, I can nevertheless conclude from my limited evidence that in the NICs too there are a number of firms which claim that the scarcity of skills is an important factor in choosing CNC lathes instead of conventional lathes.

Even in NICs that are not plagued by an absolute scarcity of skilled labour, CNC lathes should prove to be an attractive investment choice for entrepreneurs.

The above examples used Swedish cost data from 1980–81. They were also based on actual p-factors. It is of interest to analyse the investment decision in an economy with a different factor endowment. I have therefore made a sensitivity analysis as regards the choice between CNC lathes and engine lathes with different p-factors on the basis of actual cost data in Taiwan from 1983. I have used three different p-factors: 2, 3 and 4. Given the assumptions specified in Tables 2.12 and 2.13, it can be seen that, with the actual costs of the different types of lathes and the cost of labour, a p-factor of more than approximately 2 is required in order for CNC lathes to be chosen if one-shift operation is assumed. A p-factor of less than 2 would be required if two-shift operation is assumed.

The same problem can be analysed in terms of break-even wages for the skilled workers in the various alternatives. It is shown in Table 2.14 that, with a p-factor of 2, the break-even annual wage for skilled workers is 7,652 USD if CNC lathes are to be chosen, assuming the least favourable case of one shift. The present wage is about 6,600 USD p.a. For a p-factor of 3, the break-even wage drops to less than 3,000 USD. In terms of industrial reality, this would mean that if a CNC lathe oper-

Table 2.12 *Choice between CNC lathes and conventional lathes in Taiwan assuming one-shift operation[a] (US$)*

	CNC lathe	Conventional lathe		
		p = 4	p = 3	p = 2
Initial investment costs	35,000	20,000	15,000	10,000
P.a. depreciation	3,500	2,000	1,500	1,000
Interest	1,750	1,000	750	500
	5,250	3,000	2,250	1,500
Repair and maintenance costs	2,800	1,000	750	500
Labour costs				
(a) operators	6,600	26,400	19,800	13,200
(b) programming, etc.	1,320	—	—	—
Total costs	15,970	30,400	22,800	15,200

[a] Assumptions: (1) price of CNC lathe is 35,000 and of conventional lathe 5,000; (2) 10 per cent interest rate; (3) repair and maintenance costs are 8 per cent of the initial investment for CNC lathes and 5 per cent for conventional lathes; (4) labour cost is 6,600 per year; (5) 0.2 labour for programming per operator for CNC lathes; (6) one shift gives a depreciation time of 10 years.

Table 2.13 *Choice between CNC lathes and conventional lathes in Taiwan assuming two-shift operation[a] (US$)*

	CNC lathe	Conventional lathe		
		p = 4	p = 3	p = 2
Initial investment costs	35,000	20,000	15,000	10,000
P.a. depreciation	4,666	2,000	1,500	1,000
Interest	1,750	1,000	750	500
	6,416	3,000	2,250	1,500
Repair and maintenance costs	2,800	1,000	750	500
Labour costs				
(a) operators	13,200	52,800	39,600	26,400
(b) programming, etc.	2,640	—	—	—
Total costs	25,056	56,800	42,600	28,400

[a] Assumptions as in Table 2.12, except for the depreciation time which is 7½ years for CNC lathes.

ates at three times the speed of an engine lathe, and if the preparation time is 20 minutes for a CNC lathe and 10 minutes for an engine lathe, the minimum batch size would be approximately 6 units.

In other NICs, with lower wages for skilled workers (assuming same

Table 2.14 *Break-even wage for the choice of CNC lathes in Taiwan (wage p.a. in US$)*

	$p = 4$	$p = 3$	$p = 2$
One shift	1,446	2,805	7,652
Two shifts	931	1,726	4,510

Source: Tables 2.13 and 2.14.

capital costs), the *p*-factor would of course need to be raised in order for CNC lathes to be chosen. This means, *ceteris paribus*, that the scope for CNC lathes would be reduced. For example, one firm claimed in 1983 that the cost of a Korean skilled worker is about 4,900 USD per year. The *p*-factor would then, using the cost assumptions in Table 2.12, need to be about 2.4 for firms to choose CNC lathes instead of engine lathes.

I have also calculated the break-even wage using Argentinian data from 1981 (see Table 2.15). The structure of costs and the price of the dollar were very different in Argentina at this time but, if we assume two shifts, the *p*-factor above which CNC lathes would be chosen would be 2.3. In the case of one-shift work, the *p*-factor would be 2.9. The cost data are shown in Tables 2.16 and 2.17.

Table 2.15 *Break-even wage for the choice of CNC lathes in Argentina (wage p.a. in US$)*

	$p = 4$	$p = 3$	$p = 2$
One shift	5,535	10,611	28,375
Two shifts	3,541	6,509	16,895

Source: Tables 2.16 and 2.17.

Hence, in all these countries, and in particular in Taiwan and Korea, CNC lathes would be an attractive investment proposition. This is reflected in the actual data on investment in CNC lathes in these economies. On average in these three countries, an estimated 29 per cent of the value of investment in lathes was made in CNC lathes in 1980–81. This figure can be compared with an average of around 50 per cent in the developed countries in these years. The rate of diffusion of CNC lathes in these countries will be further analysed in the respective country chapters, 5, 6 and 7.

Table 2.16 *Choice between CNC lathes and conventional lathes in Argentina assuming one-shift operation[a] (US$)*

	CNC lathe	Conventional lathe		
		p = 4	p = 3	p = 2
Initial investment costs	130,000	72,000	54,000	36,000
P.a. depreciation	13,000	7,200	5,400	3,600
Interest	6,500	3,600	2,700	1,800
	19,500	10,800	8,100	5,400
Repair and maintenance costs	10,400	3,600	2,700	1,800
Labour costs				
(a) operators	12,000	48,000	36,000	24,000
(b) programming, etc.	2,400	—	—	—
Total costs	44,300	62,400	46,800	31,200

[a] Assumptions: (1) 10 per cent interest rate; (2) 10-year depreciation; (3) 8 per cent repair and maintenance costs on the initial investment for CNC lathes; 5 per cent for engine lathes.

Table 2.17 *Choice between CNC lathes and conventional lathes in Argentina assuming two-shift operation[a] (US$)*

	CNC lathe	Conventional lathe		
		p = 4	p = 3	p = 2
Initial investment costs	130,000	72,000	54,000	36,000
P.a. depreciation	17,333	7,200	5,400	3,600
Interest	6,500	3,600	2,700	1,800
	23,833	10,800	8,100	5,400
Repair and maintenance costs	10,400	3,600	2,700	1,800
Labour costs				
(a) operators	24,000	96,000	72,000	48,000
(b) programming, etc.	4,800	—	—	—
Total costs	63,033	110,400	82,800	55,200

[a] Assumptions as in Table 2.16, except for the depreciation time which is 7½ years for CNC lathes.

2.5 Conclusions

Metalcutting is one of the most fundamental processes of modern society. In the past decade it has undergone important changes as regards the technology used. Manually operated machine tools are being substituted by computer numerically controlled machine tools. This process started in the 1950s but only took off after 1975 when the microcomputer began to be used as the basis for the computer numerical control unit.

As a response to the technical and economic changes in the 1970s, the choice of technique in turning has altered and this change is reflected in a decline in the market for the simpler types of lathes that the NICs have specialized in producing. This change is in itself sufficiently alarming for the lathe producers in the NICs to contemplate adjusting and starting to produce CNC lathes too. The CNC lathe technology is, however, a technology whose operation involves the use of such a combination of factors of production that one may well argue that this technology is of as great a benefit to investors in the NICs as in the developed countries. Furthermore, the break-even wage for skilled workers, when the choice of CNC lathes becomes profitable, is fairly low, indicating that even if there is not an absolute scarcity of skilled labour the use of CNC lathes can be advantageous to investors. It is also the case that there exists a widespread diffusion of CNC lathes in the three NICs in question – Argentina, Taiwan and Korea.

What about the future? The process of substitution of CNC lathes for conventional lathes will with all likelihood continue, although at a slower pace for the developed countries. If we make the assumption that the annual growth in demand for all lathes will be 5 per cent per annum and, that by 1988, the share of engine lathes in total investment in lathes will have been reduced to 7.5 per cent from around 15 per cent in 1980, the market for engine lathes will drop from 450 million USD in 1980 to 328 million USD in 1988. At the same time it is to be expected that CNC lathes will further increase their share of the lathe market and, with the same assumption of growth in demand for all lathes, and with the additional assumption of a CNC lathe share of 80 per cent (3 per cent less than the 1984 UK and Swedish share) of total investment in lathes, the demand for CNC lathes would grow from 1,540 million USD in 1980 to 3,499 million in 1988 (see Table 2.18).

Table 2.18 *Illustrative estimate of demand for CNC lathes in the developed market economies and in some NICs*

	1980 (actual) (US$ m.)	1988 (US$ m.)	Annual rate of growth (%)
Developed market economies	1,540	3,499	10.7
Argentina, Korea (1981) and Taiwan	26	91	16.9

Sources: For the actual 1980 data for the developed market economies – UNCTAD (1982a). For Korea and Taiwan – Tables 8.10 and 7.13 respectively. For Argentina – I assumed that 56 CNC lathes were sold in 1980 (45/0.80) (see Table 6.9).

For the NICs of interest to us here – namely Argentina, Taiwan and Korea – the demand for CNC lathes was about 26 million USD in 1980–81. With the assumptions of a 10 per cent annual growth in investment in lathes until 1988 and a CNC lathe share of 50 per cent in 1988, a not unreasonable assumption given the present diffusion level and the factor-saving character of the technology, the market would grow to 91 million USD in 1988. The market for engine lathes is more difficult to speculate about as the production data do not distinguish between different types of conventional lathes. It would seem, however, if the share of CNC lathes increases to 50 per cent by 1988, that the market for engine lathes will, if not decline in absolute terms, at least grow very slowly. In any case, the domestic market for engine lathes in the three NICs is very small in relation to the developed country market; the *total* lathe market in 1980–81 was around 87 million USD in these three countries. Any marginal changes in the size of the domestic markets would not change the conclusion that the size of the world market for engine lathes is probably going to decline even further in the future. On the other hand, the market for CNC lathes will further increase, in the NICs too. Indeed, the estimated size of the market for CNC lathes in these three NICs alone would be equal to the output of six firms of the size of the largest Korean or Taiwanese CNC lathe builders today. (If we add the potential market in the remaining NICs – Brazil, Mexico, India and Singapore, which jointly have a larger machine tool consumption than the three countries studied here – the NIC market for CNC lathes would of course increase substantially.)

The conclusion from all this would be that it is strategically important for the established producers of engine lathes in the NICs to enter into the production of CNC lathes. Indeed, realizing what is happening, a number of machine tool firms in these countries, perhaps 12–15 out of about 100, are beginning to try to switch over to the production of CNC lathes. Hence, in the subsequent analysis, I shall be dealing with a few factors only. In the next two chapters I shall analyse the industry that these NIC-based firms are trying to enter.

3
Growth and Market Structure in the International CNC lathe industry

3.1 Introduction

As an introduction, it is useful to describe statistically the growth of production and trade in CNC lathes in the decade 1975–84. I have chosen to begin in 1975, not only on account of the availability of data but also because the first microcomputer-based CNC unit was produced in 1974. The importance of this technical development for the diffusion of CNC lathes was discussed in section 2.2.

In Table 2.6 we saw how the value of production of conventional and CNC lathes has altered in the main non-socialist developed countries. Together, these countries account for more than 80 per cent of the non-socialist world's production of machine tools. I noted that, in 1975, the share of CNC lathes in the total production of lathes was only 27.9 per cent whereas by 1984 it had reached 73 per cent.

In terms of the geographical distribution of production, a marked shift has taken place in favour of Japan. Table 3.1 shows that, in terms of value, the Japanese increased their share of world production from about 15 per cent in 1975 to nearly 54 per cent in 1984. Particularly noteworthy is the 14 per cent increase in 1984. As exchange rate fluctuations are common, I have also calculated the shares of Japan, Europe and the USA using the 1976 exchange rate. The Japanese share would then have been 42 per cent in 1984. In terms of units produced, Table 3.2 reveals that Japan increased its share of world production from 30.0 per cent in 1975 to 72.3 per cent in 1984.

The Europeans held their market share well, in terms of value, until 1984 when their share dropped from 42 per cent to 32 per cent, although part of this drop can be explained by exchange rate fluctuations. If we measure the value of output using the 1976 exchange rate, the European share dropped from 51 per cent in 1983 to 44 per cent in 1984. The US firms have been the main losers, however, with a reduction in their market share from 47.8 per cent in 1975 to 14 per

Table 3.1 *The production of CNC lathes in Japan, Europe[a] and the USA (in US$m. and as % of world production)*

Year	Japan				Europe[a]				USA		
	(1) US$ m.	%	(2) US$ m.	%	(1) US$ m.	%	(2) US$ m.	%	US$ m.	(1) %	(2) %
1975	66.0	15.2	66.0	14.8	156.4	35.9	166.2[b]	37.3	212.7	48.8	47.8
1976	88.7	17.8	88.7	17.8	203.2	40.8	203.2	40.8	205.9	41.3	41.3
1977	143.7	23.8	159.0	25.4	263.3	43.7	271.5	43.3	195.3	32.4	31.2
1978	194.8	24.1	274.9	29.3	373.4	46.3	425.8	45.4	237.2	29.4	25.3
1979	331.0	29.9	448.5	34.2	427.4	38.6	514.4	39.2	347.2	31.4	26.5
1980	513.7	31.8	673.0	35.3	619.9	38.3	751.7	39.4	481.0	29.8	25.2
1981	543.0	33.6	730.0	41.0	634.0	39.2	611.0	34.2	441.0	27.2	24.7
1982	473.0	33.4	563.0	39.4	617.0	43.6	539.0	37.7	325.0	23.0	22.8
1983	410.0	31.4	512.0	40.0	668.0	51.1	541.0	42.2	227.0	17.4	17.7
1984	650.0	42.1	812.0	53.8	682.0	44.2	486.0	32.2	212.0	13.7	14.0

Sources: Elaboration of data supplied by the various countries' machine tool producers associations.
(1) 1976 exchange raes.
(2) Current exchange rates.
[a] FRG, France, Italy, UK and Sweden.
[b] Excluding Italy.

Table 3.2 *The production of CNC lathes in Japan, Europe[a] and the USA (units)*

Year	Japan No.	%	Europe[a] No.	%	USA No.	%	Total
1975	1,359	30.0	1,535	33.8	1,640	36.2	4,534
1976	2,073	41.0	1,656	32.8	1,321	26.1	5,050
1977	3,900	52.6	2,332	31.5	1,178	15.9	7,410
1978	4,986	49.8	3,551	35.5	1,464	14.6	10,001
1979	8,065	57.9	3,505	25.2	2,354	16.9	13,924
1980	12,036	60.4	5,137[b]	25.8	2,751	13.8	19,924
1981	12,133	63.6	4,904	25.7	2,021	10.6	19,058
1982	10,344	64.4	4,225	26.3	1,489	9.2	16,058
1983	10,020	65.3	4,106	26.8	1,203	7.8	15,329
1984	16,555	72.3	4,818	21.0	1,524	6.7	22,897

Sources: As for Table 3.1.
[a] FRG, France, Italy, the UK and Sweden.
[b] Assuming production of 300 units in Sweden.

cent in 1984 in terms of value of production. In terms of units produced, the US firms experienced an even more spectacular loss in market shares – from 36.2 per cent in 1975 to 6.7 per cent in 1984. The Europeans' share decreased from about one-third of world production to about one-fifth between 1975 and 1984.

The large increase in Japanese production of CNC lathes has been accompanied by an export drive. In 1975, the Japanese export ratio (exports/gross output) was only 34 per cent whilst the European export ratio was well above 40 per cent. By 1981, however, the Japanese export ratio had increased to 69 per cent whilst the European had dropped to just above 30 per cent. (For one explanation of this behaviour see section 3.3.1(iii).) Since 1981, the Japanese export share has been just below 50 per cent. The European countries, taken together, have roughly the same export share, although European exports outside Europe are very small. In the case of the USA, the trade statistics only started to identify CNC lathes in 1980. In that year, the export ratio was only 4.7 per cent, and in 1981 it was only around 6 per cent. We can assume that the export ratio was also low in 1975. The total value of US production in 1975 was 212 million USD. The EEC imports of CNC lathes from the USA that year were valued at only 5.6 million USD, and Japanese imports have never been of any significant size, as will be discussed further below. It would therefore seem likely that the US producers catered mainly for the domestic market in 1975 too.

Calculating the Japanese share of the non-Japanese world market (Japanese exports divided by non-Japanese production and Japanese exports), I come to the astonishing result that it grew, in terms of units, from 12.6 per cent in 1975 to nearly 50 per cent in 1984. Thus, every other CNC lathe sold in the West in 1984 was made in Japan. In terms of value, the Japanese share rose from less than 6 per cent in 1975 to around 35 per cent in 1984. These developments can be seen in Table 3.3.

Table 3.3 *The Japanese share of the non-Japanese world market for CNC lathes (%)*

Year	Units	Value (1)	Value (2)
1975	12.6	5.7	5.6
1976	22.1	9.0	9.0
1977	29.0	13.2	14.2
1978	34.8	16.9	21.0
1979	41.7	20.9	24.3
1980	45.5	23.6	26.5
1981	48.8	28.8	35.7
1982	43.4	21.8	26.5
1983	41.1	18.3	24.6
1984	48.7	25.1	34.9

Sources: As for Table 3.1.
(1) 1976 exchange rates.
(2) Current exchange rates.

As far as the European producers are concerned, Table 3.4 shows the shares of the various countries in total European production. The most notable features are the dominance of the West German producers and the decline in Sweden's share. The UK share has increased but this is largely owing to a rise in production of very simple teaching CNC lathes since 1982.

Table 3.5 presents a summary of the main trade flows in 1975, 1980 and 1983/84. The trade in 1975 consisted mainly of exports from the EEC to Comecon countries and to the developing countries. Out of total Japanese exports in that year of 22.3 million USD, only 12.5 million USD went to the EEC and the USA. A substantial amount went to the developing countries and to countries like Australia and South Africa. Hence, the main trade flows were between technologically advanced and less technologically advanced nations or represented technological gap trade. (See, *inter alia*, Hufbauer, 1966, and Soete, 1981, for a discussion of trade based on a technological gap.)

Table 3.4 *Production of CNC lathes by individual European countries (in units and as % of total European production)*

Year	Sweden Units	%	UK Units	%	Italy Units	%	France Units	%	FRG Units	%	Total
1975	253	16	183	12	185	12	298	20	616	40	1,535
1976	238	14	185	11	300	18	259	16	674	41	1,656
1977	236	10	171	7	425	18	306	13	1,184	51	2,322
1978	242	7	288	8	600	17	547	15	1,874	53	3,551
1979	305	9	339	10	860	24	616	18	1,385	39	3,505
1980	315	6	425	8	1,552	30	650	13	2,210	43	5,152
1981	305	6	410	8	1,240	25	584	12	2,365	48	4,904
1982	184	4	573	14	852	20	617	15	1,999	47	4,225
1983	172	4	667	16	660	16	480	12	2,124	52	4,103
1984	200[a]	4	816	17	830	17	616	13	2,356	49	4,818

Sources: National and regional production data, except for Italy in 1975 where the source is Planning Research & Systems Limited (1979).
[a] Estimated.

Table 3.5 *The principal trade flows in CNC lathes in 1975, 1980 and 1983 (current US$ m.)*

Importers	United States	Exporters Japan (1984)	EEC
United States:			
1975		10.5	9.4
1980		208.5	43.1
1983		115.0 (192)	26.2
Japan:			
1975	nil		0.1
1980	0.4		1.3
1983	n.a.[a]		1.5
EEC:			
1975	5.6	2.0	
1980	6.0	128.6	
1983	n.a.[a]	60.0 (76)	
Developing countries:			
1975	n.a.	1.9	17.5
1980	6.5	18.1	34.6
1983	n.a.[a]	22.0 (51)	49.6
Socialist countries:			
1975	n.a.	nil	20.6
1980	nil	4.1	85.0
1983	n.a.[a]	17.0 (3)	178.1

Sources: Chudnovsky et al. (1983); for 1983: Eurostat, *NIMEXE Analytical Tables* for EEC exports; Japan Tariff Association (1983, 1984) for Japanese export data.
[a] Total value of exports of CNC lathes in 1983 was 13 million USD.

In 1980 and 1983/84, in contrast, the main trade flow consisted of Japanese exports to the USA and the EEC. The EEC exports to the Comecon countries continued to grow rapidly however, as did exports from the EEC and Japan to the developing countries. The trade between the technologically advanced countries has, however, become dominant over the trade from the technologically advanced nations to the developing countries. It may also be noted that there is an emerging export of CNC lathes from the NICs, mainly Taiwan and Korea, to the developed countries. In 1984, Taiwan and Korea exported CNC lathes to the US market to a value of 3.5 million USD and in 1983 the EEC imported CNC lathes from the developing countries to a value of 3.8 million USD.

GROWTH AND MARKET STRUCTURE

To summarize this introductory section, there has been a very large growth in the production of CNC lathes since 1975. In the course of this growth, Japan has become the dominant nation, at the expense of the USA in particular, but also the Europeans. The pattern of the main trade flows has shifted from one where the developed countries in the West exported CNC lathes to the technologically less advanced nations to one where Japan exports to other developed countries.

In the remaining parts of Chapter 3, I shall discuss how the changing structure of the industry producing CNC lathes, e.g. in selling concentration and in cost conditions, has altered the array of strategies actually pursued by the various firms based in the developed countries. In section 3.2, I shall present a theoretical framework for analysing the strategic position of firms within the industry. In section 3.3, I shall analyse the remarkable Japanese expansion in the production of CNC lathes. The European response to the increased competitive threat from the Japanese is discussed in section 3.4. A brief note is also made on the US CNC lathe builders in section 3.5. In section 3.6 I make some concluding remarks on the strategies pursued by the OECD-based firms.

3.2 A theoretical framework for analysing the industry

In the market model of pure competition, two main assumptions are made. First, the individual producer sells at the prevailing price in the market. Secondly, the product is homogeneous. As Chamberlain puts it: 'Goods must be perfectly homogeneous, or standardized, for if the product of any one seller is slightly different from those of others, he has a degree of control over the price of his own variety' (Chamberlain, 1960, p. 7). Often, the assumptions of free entry combined with mobility of the resources employed or potentially employed in the industry are also added in order for the competition to be *perfect* (Scherer, 1980, p. 11).

Two implications follow from the assumptions of the model of pure competition. First, as Chamberlain puts it:

> Not only goods, but sellers, must be 'standardized' under pure competition. Anything which makes buyers prefer one seller to another . . . differentiates the thing purchased to that degree, for what is bought is really a bundle of utilities, of which these things are a part. The utilities offered by all sellers to all buyers must be identical, otherwise individual sellers have a degree of control over their individual prices. Under such conditions it is evident that buyers and sellers will be paired in 'random' fashion in a large number of transactions. [Chamberlain, 1960, p. 8]

Secondly, as the market for each seller is perfectly merged with those of his rivals, the firm will not include a marketing function. The firm sells as much as it can, given the going price. Thus, in this market form, the homogeneous seller sells a homogeneous product to a homogeneous market and the decision variables for the entrepreneur are reduced to one – namely, the quantity produced.

Although Chamberlain wrote his critique of the pure competition model fifty years ago, in traditional microeconomic analysis (see for example Samuelson, 1964; Scherer, 1980) the problem focused upon is still for the entrepreneur to choose the right quantity to maximize profits. In other market forms, the price of the product is added as a second decision variable. For homogeneous oligopoly and monopoly, the quantity of output and price are the only decision variables for the entrepreneur. The product characteristics are added as a third decision variable in monopolistic competition and in differentiated oligopoly. However, although this addition is made and marketing forms are mentioned, it is the price–quantity relationship that normally receives the attention of the economist. Conventional market structure theory thus implicitly assumes that all the decision variables that influence the firm's demand and cost conditions have already been taken and focuses on the determination of the remaining decision variables – namely, price and the level of the firm's output.

Ansoff (1977) argues that this restricted focus of conventional microeconomic theory grossly limits its use in analysing the actual strategic planning process of firms. He argues (p. 12) that microeconomic theory '. . . studies the manager as an operator of a fixed arm, whereas in reality he spends much time and energy in designing the firm; its inputs, its outputs and its organisation'.

In reality, of course, neither products nor sellers are normally homogeneous, and as Chamberlain notes: 'In all of the fields where individual products have even the slightest element of uniqueness, competition bears but faint resemblance to the pure competition of a highly organised market for a homogeneous product' (Chamberlain, 1960, p. 9).

In a recent survey article, Caves (1980) reviews the concepts and theoretical framework of the positive economics of corporate strategy and compares it with that of the model of perfect competition. Caves suggests that:

> The firm rests on contractual relations that unite and coordinate various fixed assets or factors, some of them physical, others consisting of human skills, knowledge, and experience – some of them shared collectively by the managerial hierarchy. These factors are assumed to be semipermanently tied to the firm by recontracting

costs and, perhaps, market imperfections. An . . . implication of these heterogeneous fixed assets is that the firm can succeed . . . in a given market by possessing superior assets of any of several types. Equally successful market rivals thus may employ quite different bundles of fixed asset qualities. . . . The standard model of perfect competition assumes these fixed factors away, and so the concept of corporate strategy applies to market environments that could be described as imperfectly competitive. [Caves, 1980, p. 65]

Thus, Caves develops Chamberlain's (more product-oriented) notion of non-homogeneous sellers in underlining the uniqueness of each seller as the strength of his position in the market. Caves also suggests that:

Because the strategy model implies that competing firms earn different efficiency rents and that they can serve the same market by means of quite different input combinations, we expect that the products they offer to the market are multidimensional and heterogeneous, and that the firm's strategic strengths and weaknesses can be evaluated meaningfully only with respect to identified rivals. [Caves, 1980, p. 66]

When neither products, markets nor firms are homogeneous, the number of decision variables for the entrepreneur increases. These can now be listed as including:

- product characteristics
- market segment
- price
- quantity
- marketing arrangement
- R&D orientation and size.

Choosing between alternative levels of these variables and implementing that choice is referred to as the behaviour or conduct of the firm (Needham, 1978). It can also be said that the various combinations of levels of these decision variables comprise different strategies for the firm to choose between.

In reality, therefore, a firm can often influence the demand and cost conditions by choosing varying combinations of the levels of the decision variables other than price and quantities. This is true where there is extensive product differentiation within the industry, as is the case in large parts of the engineering industry. As argued by Ansoff (1977), the essence of the choice of strategy is in fact the choice of the product/

market combination of the firm, which, of course, is only of interest when the products are not homogeneous.

The degree of product differentiation in an industry is dependent on the definition of the boundaries of the industry. It is also a function of the degree of differentiability of the class of products it produces. This is largely determined by the characteristics of the buyer.

Given the definition of the industry, in the cases where there are different types of buyers, a situation may therefore prevail where the different market segments demand products that are only imperfect substitutes for one another. Choosing a particular market/product combination will however also imply that certain levels of the other decision variables need to be chosen. For example, choosing custom-designed machinery as the firm's output implies having a marketing network that can communicate directly with the customer, whereas a standardized product, in contrast, can often be sold *over the counter* by an independent distributor.

Hence, as has been argued by both Needham (1978) and Galbraith and Schendel (1983), the various decision variables are interdependent. Each set of combinations of levels of the decision variables is also associated with the use of a different mix of resources. Hence, given the restrictions of the firm, e.g. in terms of capital and skills, some combinations of levels of decision variables (strategies) may be out of its reach. Matching the firm's resources and objectives with the requirements of implementing the various strategies available then becomes the critical problem for the designer of the firm strategy. Or, as Gilmore and Brandenburg put it: 'Relating the firm's capability profile to the normative capability profile for each field or endeavour will serve to develop comparative profiles which indicate how well the firm's capabilities match the requirements for success in each field' (Gilmore and Brandenburg, 1977, p. 149).

The complexity of the market situation when each firm has chosen its particular strategy was well described by Chamberlain:

> One difficulty encountered in describing the group equilibrium is that the widest variations may exist in all respects between the different component firms. Each 'product' has distinctive features and is adapted to the tastes and needs of those who buy it. Qualitative differences lead to wide divergences in the curves of cost of production, and buyers' preferences account for a corresponding variety of demand curves, both as to shape (elasticity) and as to position (distance from the x and y axes). The result is heterogeneity of prices, and variation over a wide range in outputs (scales of production) and in profits. [Chamberlain, 1960, p. 81]

The picture that Chamberlain describes is one where the sellers within an industry vary systematically in characteristics other than size, so that the industry contains subgroups of firms with different structural characteristics. Caves and Porter (1977) take up this point in a paper that extends Bain's analysis of barriers to entry. The authors suggest that '. . . the conventional view implies that barriers to entry into the industry protect all incumbent firms as a group – a logical consequence of assuming that they are homogeneous' (Caves and Porter, 1977, p. 250), whereas in reality the barriers to entry are specific to the group. The authors also make the important point that 'barriers to mobility *between groups* rest on the same structural features as barriers to entry into any group from outside the industry' (Caves and Porter, 1977, p. 250; italics in original).

Two implications follow. First, 'each of the standard sources of entry barriers *can vary with the characteristics that define industry groups*. Entry can be easy into one group in an industry, blocked into another. Secondly, entry from outside the industry is no longer a simple yes–no choice' (Caves and Porter, 1977, p. 254; italics in original). Rather, Caves and Porter suggest that 'entry must be targeted to a particular group' and 'each of an industry's groups faces its own queue of potential entrants because of the group-specific character of entry barriers and the differing initial resources of potential entrant firms' (Caves and Porter, 1977, p. 255). It also follows that the lower the entry barriers are to a specific group (in other words, the greater the degree of similarity between the resources of the would-be entrant and those required to compete successfully within a group), the lower are the risks involved in a change in strategy by the firm (i.e. entering).

Ansoff describes the process by which a firm changes its strategy:

> A key step in the shift is the discovery of a product-market idea. However, before the idea becomes a part of the firm's product-market portfolio several other steps must be taken: enough information must be developed to convince management of the profitability of the idea, organizational competence must be developed for manufacturing, distributing, and marketing the product; recognition and acceptance by potential customers must be gained for it. *Thus strategic change is not an instantaneous event, but a protracted time and cost consuming process.* [Ansoff, 1977, p. 21; emphasis added]

Thus, the first step in the process of strategic change is the *search* for a novel product/market combination, which is usually both costly and time consuming. An important factor determining the *scope* of the search is the affinity between the various search areas and the present position of the firm. Naturally, the greater the affinity (manifesting itself, for example, in the use of the same skills in the search areas as in the present position), the lower are the risks involved in a move. Another influential factor determining the scope of the search is the critical mass (e.g. financial and technological) perceived by the firm as being necessary for a successful entry into the new area. If a firm perceives the critical mass to be substantially greater than the firm's means, a search will not take place. In other words, certain strategic choices will not be explored in detail by the management.

After a search process has identified a potential new strategic orientation, the management proceeds to develop proof of the profitability of the idea. As Ansoff explains:

> ... at the outset the estimates of profitability are usually very rough. Information is lacking about what it will take to develop the needed capabilities, what resources must be committed, how many units will sell, at what prices, etc. ... as the process progresses, the information about profitability improves and uncertainties are reduced, but at the same time, investment rises at an increasing rate ... When a point is reached at which management is prepared to commit itself to the new product-market, the major emphasis is shifted to development of organizational skills and capabilities for producing and marketing the product ... During this period full-scale manufacturing facilities are built, market and distribution channels are established ... [Ansoff, 1977, pp. 23, 24]

Ansoff distinguishes between an *expansion type* strategic change and a *planned diversification* type of strategic change. The former strategic change involves large similarities between the resources used for the present strategy and the resources required for the new strategy. An expansion type strategic change is often a natural part of a firm's dynamics where, for example, market shares are taken from competitors, or new markets abroad are conquered, using broadly the same product/market combination. The rate of *strategic expansion* (Ansoff, 1977, p. 30) is determined by the budgets allocated to such variables as R&D, marketing capabilities and production capacity. The risks involved in such a change are those of a possible surplus capacity and those associated with developing new markets, particularly abroad. A *planned diversification*, on the other hand, involves a new product/market combination for the firm, which often requires redesigning and reconstituting the firm (Ansoff, 1977, p. 32) and therefore new organization and skills. Naturally, the risks involved in a planned diversification can be perceived as much higher than those of a strategic expansion.

3.2.1 The industry in a dynamic context

The problem of choosing and changing strategy for the firm is further complicated by the fact that the initial conditions determining the structure of the industry may alter as a consequence of, for example, technical change. Hence, the *menu* of different viable strategies, or the nature of competition in the industry, may well vary over time.

It is common to describe the course of industry evolution in terms of the life cycle of the product of the industry. The product/industry is seen to pass through a number of phases – introductory, growth and maturity phases. A decline phase is also usually included. Figure 3.1 plots the real S-curves of CNC lathes in Sweden, Japan and the UK. In terms of this terminology, CNC lathes entered their growth phase around 1976 in the case of Japan and Sweden and in 1977 in the case of the UK. (Apart from the case of the UK, they do not resemble so well the S-curves in the textbooks.) Many attempts have been made to use the S-curve concept as a basis for predicting how an industry changes over its life cycle and how these changes affect company strategies (Porter, 1980). The problem with using the concept for predictive purposes is simply that it is an attempt to describe *one* pattern that is thought to be applicable to all industries/products. Whilst I recognize the limitations of the concept, I have nevertheless found it to be a useful tool in my analysis of the CNC lathe industry. The subsequent discussion should therefore not be seen as one where I make very general statements of how industries change in the course of their pro-

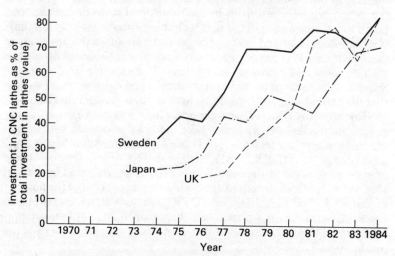

Figure 3.1 *The S-curve for CNC lathes in Sweden, Japan and the UK*
Source: Table 2.5.

ducts' life cycles. Ample references will instead be made to the case of NCMTs.

As an industry evolves or moves along its S-curve, new market segments are penetrated. The change in market size as well as in the segments penetrated can be a function of a number of factors. Common ones are product innovations that improve the performance or reliability of the product, product differentiation, improved knowledge of the technology among potential users, changing factor prices, etc.

In the first stage of the product cycle, the product is introduced. It is often produced in small quantities and for very specific customers. Utterback suggests that: 'The initial uses of a major product innovation tend to be in small . . . market niches in which the superiority of the new product . . . allows it to command a temporary monopoly, high prices and high profit margins per unit' (Utterback, 1979, p. 47). Often, the new product is characterized by functional superiority rather than by a lower price. In this phase, close contact with the market is also necessary for the producer in order to acquire ideas for product developments.

The diffusion of the new technology to other users is generally a very lengthy process. When a new product, a machine, first comes on to the market, it is often technically immature and very costly in comparison with the machine it will eventually replace. Furthermore, knowledge about the new technology is generally lacking among potential customers. In the case of NCMTs, a study in the early 1970s showed that the difference between firms in the developed countries in the time taken to become aware of the new technology was as great as ten years (Gebhardt and Hatzold, 1974). Larger firms are often the first users of the new technology as they generally have more inhouse skills to deal with the new and immature technology. For example, in the case of Sweden, first users of NCMTs often bought one to experiment with so that they could see how it could be used in their production (Elsässer, 1983b). Moreover, larger firms generally have more information about new technologies and can better take the risks associated with investing in an unproven technology. These large firms often become the *leading edge firms* (DEK, 1981). In the case of Sweden, the dominance of such large firms in the early use of NCMTs is reflected in the fact that, in 1976, fifteen firms had approximately 60 per cent of the number of installed CNC machine tools (DEK, 1981). Watanabe (1983) also reports that, in the early phases of diffusion of NCMTs in Japan, larger firms (more than 300 employees) constituted a very large share of the market (see Table 3.6).

As the product moves to its growth phase, changes take place both in the type of users of the new machine and in the machine itself. Outside

Table 3.6 *Investment in NCMTs in large and small enterprises in Japan, 1970–1981 (yen m.)*

year	Large firms	Medium and small firms[a]
1970	15,510	7,404
1971	17,278	8,639
1972	13,951	10,600
1973	23,075	25,122
1974	25,310	25,547
1975	12,921	17,756
1976	17,069	22,178
1977	23,820	24,856
1978	19,957	38,445
1979	49,013	79,017
1980	68,847	126,960
1981	92,068	153,292

Source: Elaboration of Watanabe (1983).
[a] Firms with fewer than 300 employees.

the large product-oriented firms, for example ASEA and VOLVO in Sweden, there are a large number of medium-sized and smaller firms that produce general industrial machinery and do subcontracting work. These firms often demand a different type of technology from the larger firms. Generally speaking, these smaller firms have less inhouse technical expertise and are therefore forced to rely more on mature technologies. Because they are smaller, they cannot in general profitably use as much specialized machinery as the larger firms but need more universal and standardized machine tools; they are more price conscious. Their smallness also means that scale economies in buying the new technology (in terms of information search) and in using it (in terms of the need for specialized skills, for example programmers) have to be reduced before they can buy the new technology.

If the product/industry is to move along its S-curve, the product needs to be altered in order to satisfy the differing demands made by the two broad groups of customers. The growth phase can thus be characterized by the penetration of new markets through:

- the standardization of the technology, i.e. fewer custom-designed features;
- product differentiation (away from the specifications set by the larger customers) broadening the range of models available to customers to include more *simple* types;

- price reductions as a consequence of both product differentiation into more *simple* models and the achievement of economies of scale.

Table 3.6 shows how a growing part of the Japanese market for NCMTs was constituted by firms with fewer than 300 employed; it is also reported that, in 1982, 26.1 per cent of the total value of sales of NCMTs in the Japanese market went to firms with fewer than thirty employees (Watanabe, 1983; see also *Metalworking, Engineering and Marketing*, 1983a). In a study of the diffusion of NCMTs in the non-electrical machinery sector (ISIC 382) in Sweden, Elsässer and Lindvall (1984) found that twenty-nine out of the forty-four firms that had acquired their first NCMT before 1969 had more than 500 employees. On the other hand, of the twenty-eight workplaces that acquired their first NCMT between 1975 and 1980, twenty-one had fewer than 500 employees. Hence, the smaller and medium-sized firms now constitute a large part of the market for NCMTs, including CNC lathes, and their share of the market has grown with the increasing maturity of the technology of NCMTs.

A change in the size and structure of the market is often reflected in a change in the industry supplying the product (Porter, 1980). As new requirements have to be met and production increases in volume, the original conditions determining the structure of the industry may alter and, with them, the range of strategies available for firms to choose. Of course, a change in the structure of the industry is not independent of the individual firm's strategy and the actual definition of the S-curve of a particular product/industry is the combined result of the effects of a number of firms' strategies.

In summary, the strategic position of the CNC lathe producers in the introductory phase of the diffusion of CNC lathes can be characterized as follows:

- performance rather than price mattering in the design of the product;
- a close relationship with larger firms, which demanded high performance and often custom-designed lathes, being the rule;
- a heavy reliance on product development as a competitive tool;
- low volume of output of CNC lathes per firm;
- a tendency for CNC lathe producers to sell directly to the user rather than through a distributor and for the user to have sophisticated questions about engineering problems.

A case in point here is the UK firm, T. I. Churchill, which until very recently relied on producing large and powerful CNC lathes, often custom-designed, to large firms in the transport industry (interview).

Furthermore, the trade patterns shown in Table 3.5 suggest that the business relations were mainly of a local or regional nature, the exceptions being EEC exports to the Comecon countries and exports from the developed countries to the developing countries. Among the three OECD regions, no major differences with respect to technological knowledge seem to have existed. Hence, many of the characteristics of a product in the early part of its life cycle applied to CNC lathes at this time. Production was also fairly evenly distributed between the USA, Europe and Japan. In 1973, for example, US production was around 1,000 units, Japan's was 1,459 units and the FRG's was 501 units (NMTBA, 1981/82; JMTBA, 1980; data from VDW).

When the CNC lathe industry moved to its growth phase in the mid-1970s, the structure of the industry was greatly affected. A new market of medium and smaller firms was opened up, which required a fundamental shift in strategy among some suppliers. In the following sections I shall discuss how some Japanese firms understood the significance of opening up this market and how, by implementing a new strategy, they changed the rules of the game in the international industry.

3.3 The Japanese expansion in the CNC lathe industry

In the mid-1970s, some Japanese firms started to apply a strategy that could be labelled an 'overall cost leadership' strategy. The firms had as a basic objective the penetration of a very large part of the engineering industry. This was achieved by: differentiating their CNC lathes from those traditionally supplied in the other OECD countries; a standardization of the CNC lathes; an increase in the production volumes, which thereby enabled them to benefit from various sources of economies of scale. In terms of the S-curve terminology, these firms created the point of the maximum curvature and made the industry move upwards on the S-curve.

The key factor in the definition of this strategy was the design of lower-performance, smaller and lower-cost CNC lathes. The target market for these lower-performance and standardized CNC lathes was primarily the smaller and medium-sized firms. In other words, these Japanese firms deliberately and successfully, set out to open up the large market of smaller and medium sized firms, which, as I argued above, often demand a slightly different type of technology from larger firms. The European and US producers, on the other hand, very broadly speaking, took the production problems in larger firms as the point of departure for their design efforts.

The differences between the types of CNC lathes produced in Japan and in other OECD countries can be illustrated by three sets of data. First, weight per unit is an indicator not only of the size of work piece that can be turned but also of the power that can be applied to the work piece. The lower the weight, the lower the performance that can be achieved with the lathe. Table 3.7 shows the differences between Japan and the Federal Republic of Germany with respect to the weight in tons per unit of the CNC lathes produced. Secondly, the motor power is a determinant of the performance of the lathe, i.e. the cutting speed and power. Of the CNC lathe models made in Japan in 1979, 38 per cent had a motor power of less than 15 kW, another 38 per cent had between 15–30 kW and only 24 per cent had a motor power greater than 30 kW (Jacobsson, 1982a). On average, this is substantially lower than most European and US firms. For example, the leading Swedish firm, which is one of the five largest European firms, has no model with a motor power of less than 35 kW (firm interview). Thirdly, the price differences between CNC lathes produced in Japan, the USA and the Federal Republic of Germany (shown in Table 2.3), strongly suggest that the Japanese CNC lathes are of lower performance.

If we look at the market profile in the US, which is the world's largest market, it is also clear that the Japanese firms chose to develop a market segment of substantial size. In 1980, CNC lathes with a motor power of less than 18 kW accounted for 46 per cent of the market in terms of value and over 62 per cent in terms of units. This can be seen from Table 3.8. In 1983, the share of smaller CNC lathes had increased to 54 per cent and 75 per cent respectively (NMTBA, 1984/85).

Table 3.7 *Weight per CNC lathe produced in Japan and in the FRG (tons)*

Year	Japan	FRG
1973	5.5	13.0
1974	5.8	12.9
1975	5.0	13.4
1976	4.9	13.7
1977	4.1	9.5
1978	4.0	7.4
1979	4.3	10.2
1980	4.6	9.2
1981	4.7	7.7
1982	4.8	8.2
1983	4.1	8.9
1984	4.2	8.4

Sources: Japan: 1973–80 data supplied by JETRO; 1981–84 data supplied by CECIMO. FRG: elaboration of data supplied by VDW and CECIMO (for 1984).

Table 3.8 *Market profile for horizontal CNC lathes in the USA with respect to the strength of the motor, 1980*

Market size	Less than 18 kW	Motor strength 18–37 kW	More than 37 kW
Units	2,899	1,422	312
Value (US$ m.)	249	220	71

Source: Elaboration of NMTBA (1981/82).

The different philosophies with respect to the design of the CNC lathe are well illustrated in a Japanese journal article of 1977:

> At the first stage, these low cost NC lathes could not be introduced smoothly due to the prejudice that NC lathes should be machine tools of high quality equipped with luxurious functions. Though NC lathes of high grade with luxurious functions were really required for some turning operations, it is also true that all the valuable functions are not required for all the turning operations. In many fields, NC lathes of simplified functions can sufficiently turn the parts, and many low-cost NC lathes are now accepted positively. [*Today's Machine Tool Industry*, 1977a, pp. 70–2]

Thus, broadly speaking and using the concept of a performance/cost ratio, it may be said that the largest Japanese firms have emphasized simplification and thereby cost reduction in their design efforts, whilst the majority of European and US firms have emphasized performance. The widening of the spectrum of choice of CNC lathes has also clearly had the effect of opening up new markets and has thereby been conducive to an increase in the scale of output. The changing nature of the CNC lathes produced and the increases in the scale of output, which led to reaping economies of scale, are also shown by the fact that the price ratio of units purchased of CNC lathes and conventional lathes dropped in Japan from 8.3 in 1974 to 2.9 in 1981 (see Table 2.4).

The leading Japanese firms have increased their rates of output in a way that is phenomenal in the context of this industry. Table 3.9 shows that the period after 1975 was characterized by a very rapid increase in output by the five largest firms. These firms held, and indeed slightly increased, their share of both the value of output and the units produced: in 1981, the five largest firms accounted for 76 per cent of the value of CNC lathes produced and 66 per cent of the units.

Table 3.9 *Concentration of production of CNC lathes in Japan, 1970–1981 (in value, units and %)*

	1970	1975	1978	1979	1980	1981
Total Japanese production of CNC lathes:						
Value (current yen m.)	8,893	19,618	57,862	98,320	152,443	161,287
Units	589	1,355	4,986	8,203	12,007	12,133
The production of the largest firm and its share of total Japanese production:						
Value (current yen m.)	3,557	3,727	10,993	20,549	31,708	38,709
%	40	19	19	21	21	24
Units	236[a]	257[a]	950[a]	1,567	2,173	2,438
%	40[a]	19[a]	19[a]	19	18	20
Production of the largest five firms and their share of total Japanese production:						
Value (current yen m.)	8,181	11,770	37,031	64,202	92,837	122,416
%	92	60	64	65	61	76
Units	541[a]	813[a]	3,191[a]	5,003	6,723	8,056
%	92[a]	60[a]	64[a]	61	56	66

Sources: Elaboration of data supplied by the JMTBA.
[a] Estimated.

Hence, in terms of the six decision variables discussed in section 3.2, the Japanese firms

- simplified the product, i.e. altered the product characteristics;
- aimed for smaller and medium-sized firms, i.e. altered the market;
- drastically increased the volume of output;
- emphasized price as a competitive tool to a greater extent than had hitherto been the case;
- increased the proportion of indirect sales, i.e. *over the counter* sales, through distributors rather than direct sales to the customer;
- directed R&D activities to satisfy the needs of medium to small firms.

This strategy has also been extremely successful for most of these companies in terms of both growth rates and financial returns. Data on sales and profits are shown in Table 3.10 for five of the leading firms. On the whole, the implementation of the strategy of overall cost leadership in CNC lathes since 1975 has been associated with, and is probably the main reason for, the very large profits won by these five firms after 1979. The Boston Consulting Group also concludes: 'The rewards of the successful volume strategy are a defensible market position and a high profit potential . . . The returns on sales of leading Japanese volume suppliers show levels of profitability which remain

Table 3.10 Sales and profits[a] of five Japanese CNC lathe producers[b]

Firm		1975	1976	1977	1978	1979	1980	1981	1982	1983	1984	1985
(1)	Sales (yen b.)	5.8	10.6	13.3	17.0	23.3	37.9	53.3	57.2	47.1	46.5	69.5
	Profits (%)	−14.3	−25.4	−11.9	−7.5	5.2	15.4	16.5	15.9	8.5	6.5	10.9
(2)	Sales (yen b.)	8.3	9.7	12.7	15.6	18.5	26.1	33.7	38.6	30.6	30.0	40.0
	Profits (%)	0.9	−23.0	−18.9	−8.2	−5.4	6.6	11.8	10.9	3.1	0.0	3.0
(3)	Sales (yen b.)	15.4	12.2	16.3	19.6	24.2	27.2	29.7	26.8	22.2	24.1	25.0
	Profits (%)	−5.6	−22.1	−14.9	−0.5	1.9	2.6	1.8	−2.7	−12.1	−9.9	4.8
(4)	Sales (yen b.)					18.5	27.5	37.9	45.4	25.3	34.3	52.0
	Profits (%)					23.3	27.6	31.7	31.6	15.4	10.6	19.8
(5)	Sales (yen b.)				17	25	43	53	60	n.a.	n.a.	n.a.
	Profits, average (1978–82) 8–12% of sales											

Sources: The Oriental Economist, Japan Company Handbook (various years); and an interview with a firm in Japan.
[a] As % of sales as of March each year.
[b] The data cover the total output of the firms. These are, however, specialized machine tool builders, concentrating on CNC lathes in their output at present.

above the average even in years of worldwide depression in the industry' (Boston Consulting Group, 1985, p. 30).

The implementation of this strategy by these firms has had the additional effect of changing the structure of the industry. Seller concentration has increased, although marginally, and economies of scale are far more important today than in the middle of the 1970s. The average size of the five leading firms has increased dramatically, both in Japan and in Europe. These issues are discussed in greater detail in sections 3.4 and 3.5, where I shall also briefly discuss vertical integration.

3.3.1 A discussion of the factors behind the Japanese success

What are then the reasons behind the Japanese success? Without pretending to be able to answer this question fully, I shall in the following sections discuss some factors that may be of relevance: the character of the home market; the external economies between machine tool producers and the electronics industry; the size and closed character of the domestic market; and the financial aspects of strategic choices.

(i) The character of the home market in Japan The preceding discussion emphasized the importance of product differentiation in the Japanese success. It would therefore be natural to ask if the Japanese home market is different with respect to the composition of large and small firms from that in other countries. Watanabe (1983) discusses this issue in detail, emphasizing the importance of the market for capital goods that the smaller firms constitute in Japan. The *unique part of Japanese industrial organization*, the subcontracting system, is particularly concentrated on:

> ... the Japanese industries are characterized by the existence of vast numbers of medium and small enterprises and by their cooperative operation with larger firms as their subcontractors. Given this industrial structure, the capital goods manufacturers cannot neglect their needs, because collectively those small enterprises provide them with an even larger market for their products than larger firms. [Watanabe, 1983, p. 71]

If correct, this argument would fit nicely into the scheme described above of the role of medium and small enterprises in the process of diffusion of NCMTs. However, there are a number of problems in the presentation of the evidence. Whilst it seems clear that the Japanese industrial structure (evidence is not given for the metalworking sector) is more fragmented than in the USA for example, Watanabe presents no good evidence that the smaller and medium-sized firms in Japan did in fact constitute a larger market for NCMTs at an earlier stage than their counterparts in the USA or in Europe. The only evidence runs as follows:

> In 1981, two-thirds of the total shipments of NC-machine tools within the country were directed to the former [the small and medium-sized enterprises]. This . . . contrasts sharply with the situation in the United States, where such machinery has been used mostly by larger firms: two-thirds of the total of 50,100 units were being used at plants with more than 100 workers. [Watanabe, 1983, pp. 22–3]

Watanabe then suggests that the difference arises largely from the existence of a vast number of small enterprises (establishments) in the Japanese industries (Watanabe, 1983, p. 22). The problems with the argument are as follows.

First, flow is compared with stock: the US figures refer to the stock of NCMTs, while the Japanese figures refer to the flow. As we know, the smaller firms tend to be late users of the technology. It is therefore misleading to compare flow and stock. Secondly, in the US case, the smaller firms are defined as those with less than 100 employees whilst in the Japanese data, the limit is 300 employees. The difference can be very serious because a large share of the value added in industry is created in firms with 100–300 employees. In the Japanese case, 16.5 per cent of value added in industry in 1979 was created in firms with 200–300 employees (elaboration of Watanabe, 1983, p. 22). Thirdly, the US data were collected during the period 1976–78, which is the time when NCMTs had just started to be mass produced and were more suited to the needs of the smaller companies. The Japanese data refer instead to 1981.

If, for the sake of the argument, we convert the flow data in Table 3.6 to stock data by adding the investment in NCMTs in Japan from 1970 to 1978 (using 1975 prices) and divide the investment between larger and smaller enterprises, the result is that larger firms accounted for 50 per cent of the stock by 1978 – compared to two-thirds in the case of the USA as regards the share of larger firms in the demand for NCMTs. However, we need also to compensate for the differences in definitions of larger firms. It is not possible to do this in a strict way as we do not know the share of the firms in size groups 100–300 employees in the demand for NCMTs in Japan. But if 16.5 per cent of the value added in Japanese industry is created in firms with 200–300 employees, it would not be unreasonable to expect that at least 15 per cent of the total stock of NCMTs in Japan is found in firms with 100–300 employees.

Furthermore, a study of the distribution of the stock of NCMTs by size of firm in the FRG in 1980 shows, as Table 3.11 reveals, that approximately 46 per cent of the stock of NCMTs was in firms with fewer than 250 employees. This is roughly comparable to the Japanese data if we put all the sales (using 1975 prices) from 1970 to 1979 into two categories: firms with fewer than 300 employees and firms with more than 300 employees. We would then come to the result that larger firms had bought about 47 per cent of the NCMTs, in value terms, and smaller firms 53 per cent. There could be a problem in that the German study measured the number of machine tools bought and the Japanese figures refer to the value of machines bought. On the other hand, larger firms in Germany include firms with 250 employees, whilst the limit is set at 300 in Japan.

Hence, it is not likely that the differences in the distribution of NCMTs between larger and smaller enterprises were so large between Japan and other countries at the end of the 1970s. This conclusion is strengthened by Steen's (1976) observation that, in 1973, 28 per cent of the NCMTs installed in the USA were in firms with less than 100 employees, whilst the figure was only 3 per cent in Sweden. Of course, this does not necessarily mean that the Japanese producers of NCMTs were not guided in their design development by the demands of smaller and medium-sized enterprises. It only means that it was not necessarily the unique feature of the metalworking industries in Japan that caused the attention of the Japanese supplier to be focused on the requirements of smaller and medium-sized firms. It may simply have been an awareness, for other reasons, of the size of this market, not only in Japan but also in other countries.

One partial explanation for this is possibly the absence of a military sector in Japan. Watanabe (1983) stresses the absence of an aerospace industry in Japan and underlines the role of the automobile industry, including makers of components, for the technological development of machine tools. This argument is in line with that of Noble (1978), who

Table 3.11 *The distribution of the stock of NCMTs by size group of firms in the FRG in 1980*

Size group of firms	No. of NCMTs	% of total stock
10– 40 employees	3,800	15.2 ⎫
50– 99	3,800	15.2 ⎬ 46.4
100–249	4,000	16.0 ⎭
250–999	5,400	21.6
1000+	8,000	32.0

Source: Data supplied by VDW.

suggests that the demand by the military sector in the USA has created a design philosophy among producers of capital goods that results in the building of *super machines*, which are of little interest to the vast majority of metalworking firms. After explaining how the US Air Force paid for the development of NCMTs and created a market for them, Noble continues:

> The point of this story is that the same thing that made NC possible, massive Air Force support, also quite possibly determined the shape the technology would take. Criteria for design of machinery normally include cost to the user. Here cost was hardly a major consideration; machine tool builders were simply competing to meet performance specifications for government funded users in the aircraft industry. They had little concern with cost effectiveness and absolutely no incentive to produce less expensive machinery for the commercial market. [Noble, 1978, p. 329]

Bhattacharay gives more evidence on the guiding role of the aerospace industry in the design development of the machine tool industry, this time in Great Britain. In an evaluation of the penetration and utilization of NCMTs in British industry in 1976, he comes to the conclusion that: 'Research and development in this country is too much concentrated on the sophisticated end of the market, which is primarily catering for the minor aerospace industry' (Bhattacharay, 1976, p. 15). He goes on: 'The major bread and butter end of the market is the simple two to three axis types of machines. It is in this area the government and machine tool companies must concentrate their research efforts. This is likely to be micro-processor control with a minimum degree of sophistication' (Bhattacharay, 1976, p. 19).

There therefore appears to be some evidence of the negative effect that the very sophisticated demand of the military sector and the civilian aerospace sector has had on the large-scale commercialization of UK and US made NCMTs. This view is further strengthened by the fact that a senior Japanese designer in one of the pioneering firms in CNC lathes claimed that when his firm began producing CNC lathes in 1967 they produced standard and not custom-designed machines. The reasons for producing standard machines lay in a desire to reduce costs and improve the position of CNC lathes vis-à-vis conventional lathes (interview). The same designer also claimed that he could not understand why the European and US firms had such powerful machines because if you put in a more powerful motor, all other parts have to be strengthened and that negatively affects the cost/performance ratio. Thus the situation in Japan would seem to have been distinctly different from that painted by Noble of the USA.

(ii) External economies between the machine tool industry and the electronics industry Of critical importance for the development of low-cost lathes was the evolution of numerical control technology. This was discussed in section 2.2. The development of cheap CNC units[1] was a prerequisite for the diversification of CNC lathe designs into the less complex part of the spectrum; it would have been very uneconomic to apply the old control technology to the relatively cheap lathes (that is, compared to milling machines, etc.). As the Japanese said themselves in 1977: '. . . the tendency of diversification [of models of NC lathes] . . . became notable while the cost reduction in NCs has been realized owing to electronic technique innovations, thus making it possible to develop inexpensive NC machine tools' (*Today's Machine Tool Industry*, 1977b, pp. 68–70).

Historically, there have been strong design links between the producers of the CNC unit and the machine tool builders in Japan. Fujitsu Fanuc, the largest producer of CNC units in the world, began its collaboration with machine tool firms in 1958 when, together with Makino Milling Machine Co., it developed the first Japanese NCMT[2] (*Today's Machine Tool Industry*, 1976, p. 53).

As can be seen in Figure 3.2, Fanuc also had an early start producing NC units and had by 1974 accumulated considerable experience of production and the demands of its customers. Then, in 1975, Fanuc succeeded in realizing a substantial reduction in the cost of its CNC

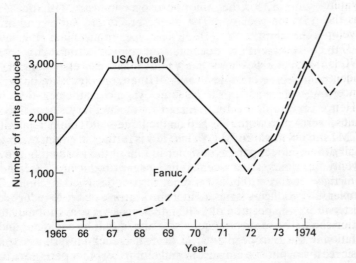

Figure 3.2. *Yearly shipment of numerical control units by US manufacturers and by Fanuc, 1964–1974*

Source: *Today's Machine Tool Industry* (1976), p. 58

units as it switched from minicomputer-based and software variable type CNC to fixed software type microcomputer-based CNC units (*Today's Machine Tool Industry*, 1976, p. 60). The differences between these CNC units were described in Chapter 2.2. Hence, the Japanese supplier of CNC units initiated the development of simpler and much cheaper CNC units, which could advantageously be applied to smaller lathes.

According to one source (*Today's Machine Tool Industry*, 1976), Fanuc has *always* emphasized low cost in its CNC units. It claims: 'In Japan, in its cradle stage, efforts were made to offer low cost CNC's to customers, even limiting NC functions to some extent so as to promote their marketing prevalence at the earliest possible time' (*Today's Machine Tool Industry*, 1976, p. 39).

In an atomized industry like the machine tool industry, the geographical and cultural proximity of producers of new components and the machine tool builders is probably very important for a rapid adoption of new types of components on account of the relatively high costs of information search for smaller firms. The strong design links between the CNC system producer and the machine tool firm would also have given the Japanese machine tool firms an advantage vis-à-vis their European and US competitors. Watanabe reports that Fanuc – the world-dominant Japanese producer of CNC units – still has regular meetings with its major users to exchange views on technical matters (Watanabe, 1983, p. 37). Watanabe also hints at the possibility that it was the machine tool producers who pushed the CNC supplier to develop cheaper and more reliable CNC units (Watanabe, 1983, p. 37). This seems to be highly likely given that the Japanese-made CNC lathes have always had a very low tonnage per unit and given the evidence of an early orientation towards larger markets and subsequent concern with the cost of the NCMTs.

Hence, there were strong complementaries, or external economies,[3] reaped from the design developments in the NC unit and in NCMTs. However, producers of CNC systems exist in large numbers in all developed countries: at the EMO exhibition in Milan in 1979, seventy-five different producers of control systems were represented (*American Machinist*, 1979). The early development of smaller and simpler CNC lathes in Japan cannot therefore be said to depend on the existence of a Japanese electronics industry *per se*; rather an explanation should be sought in the factors behind this industry's early orientation to low-cost CNC units. It would seem arguable that the more basic reasons are those connected to the character of the home market, particularly as Fanuc has always emphasized lower-cost CNC units – even before the introduction of microprocessor-based CNC units. Nevertheless, the development of cheap microprocessor-based CNC units in Japan meant that Japanese machine tool builders were given, or created for themselves, an advantage that they were quick to exploit through rapidly expanding the output of low-cost CNC lathes in 1976/77 (*Today's Machine Tool Industry*, 1977a, p. 70). It is also of interest here that the advantage was prolonged to some extent because CNC units made by Fanuc were distributed in Europe by Siemens at a much higher price than that prevailing in Japan. According to one source, Fanuc did not start to sell directly to European machine tool firms until 1983 (interview with a European machine tool firm).[4]

(iii) The size and degree of captiveness of the home market Table 3.12 compares the size of the domestic markets for CNC lathes in USA, Japan and the four largest EEC countries. In terms of units demanded, the regions are very similar, with the exceptions of the extremely large Japanese domestic market in 1984 and the depressed US market of 1983. In terms of value, the Japanese market has grown to equal that of the USA and EEC and is substantially larger than the markets of any of the EEC countries.

Whilst the US and European markets are heavily penetrated by external competitors, the Japanese home market is satisfied by only Japanese-made CNC lathes. Table 3.13 shows the trade in CNC lathes in Japan from 1969 to 1984. Japan has not imported any noticeable numbers of CNC lathes, at least not since 1969; the maximum number of imported CNC lathes was sixty-one units in 1983.

Even though I argued in section 3.1 that there was little trade between Europe, USA and Japan in the early and mid 1970s, there were still some imports into the USA around 1975. As was shown in Table 3.5, about 20 million USD worth of CNC lathes were imported into the USA from Japan and Europe in 1975; this represented approximately 10 per cent of the demand for CNC lathes. Furthermore, if we look at the individual EEC countries, the picture is very different. The import ratio for the FRG in 1976 was 34.8 per cent, in the case of France was 33.9 per cent and in the case of the UK was as high as 76.1 per cent.

Of course, given the large size of the domestic market in Japan and the fact that it was, and is, captive in a descriptive sense, the risks involved in an expansion of production, such as several Japanese firms undertook in the second half of the 1970s, could be considerably reduced. One of the leading Japanese firms, whose production in 1977 consisted of only 20 per cent NCMTs, decided to build a new plant after four years of consecutive losses. The output capacity of the new plant was close to the total demand for CNC lathes in Japan *at the time when the investment decision was taken* (interview). By 1981 they had not only increased sales by 400 per cent but also reached a 90 per cent

Table 3.12 *The size of the domestic market for CNC lathes in the USA, Japan, UK, Italy, France and FRG, 1976, 1980, 1983 and 1984 (in value and units)*

	Value (US$ m.)							Units						
Year	USA	Japan	UK	Italy	France	FRG	Total 4 EEC	USA	Japan	UK	Italy	France	FRG	Total 4 EEC
1976	206[a]	48	23	18	39	24	104	1,321	1,202	222	286	344	730	1,582
1980	589	227	132	139	80	258	609	5,001	5,427	1,126	1,309[b]	896	2,316	5,647
1983	261	263	65	51	79	n.a.	331	2,581	6,376	1,297	494[b]	898	1,663	4,352
1984	395	438	90	n.a.	93	140	n.a.	4,575	10,551	1,449	n.a.	1,001	1,661	n.a.

[a] Production.
[b] Estimated.

Sources: 1976 and 1980: US data – NMTBA (1981) and (1980); Japanese data – JETRO; UK data – MTTA; German and Italian data on apparent consumption of units in 1976 – Planning Research Systems Limited (1979); German data on units consumed in 1980 – VDW for production data and Statistisches Bundesamt (1981) for the trade data; German data on apparent consumption in terms of value – VDW for production data and Eurostat; *NIMEXE Analytical Tables* for the trade data; Italian data on apparent consumption in value in 1976 and 1980 – UCIMU for production data and *NIMEXE Analytical Tables* for the trade data; French data – Syndicat des Constructeurs Français de Machines Outils, *Statistiques commerce extérieur* (1976, 1980) and *Enquête syndicale sur la commande numérique dans l'industrie de la Machine Outil* (1977, 1981). 1983 and 1984: USA, Japan and France – NMTBA (1984/85, 1985/86); UK data – MTTA (1985); German data – VDW; Italian data – estimated on the basis of data given by CECIMO (for production) and Eurostat, *NIMEXE Analytical Tables* (1983).

Table 3.13 *Japanese trade in CNC lathes, 1969–1984 (units)*

Year	Imports	Exports
1969	14	34
1970	12	22
1971	24	31
1972	6	81
1973	7	103
1974	14	262
1975	0	456
1976	3	847
1977	12	1,436
1978	5	2,675
1979	3	4,194
1980	12	6,592
1981	n.a.	n.a.
1982	17	4,382
1983	61	3,705
1984	13	6,012

Sources: JETRO; NMTBA (1983/84 and 1984/85); *Metalworking, Engineering and Marketing* (1985); Japan Tariff Association (1985).

share of NCMTs in total production. When their expansion and specialization were decided upon, they planned for an export share of 40 per cent, but this turned out to be only 30 per cent. Furthermore, the firm expanded its production phenomenally: in 1981 it produced roughly the same amount of CNC lathes as the total demand in Japan in 1977. It is difficult to conceive of a firm in another country, with a smaller domestic market penetrated by imports, behaving in a similar fashion.

The reasons behind the near-absence of imports to Japan are not obvious. It could of course be argued that the very marginal trade between the three main regions around 1975, meant that Europe and the USA did not have a significant import share either. Table 3.12 showed, however, that the Japanese domestic market was not very different from that of the FRG (in terms of units) and that of France (in terms of value). Both these countries had a large trade in CNC lathes around 1976 (see Table 3.14). To the extent that this trade was based on a specialization in different types of CNC lathes, it would seem a bit surprising that the Japanese did not participate in such trade.

However, what is more important is the continuing near-absence of imports to Japan in the second half of the 1970s when their exports rose so dramatically. It is of interest here to mention the 'technology gap' account of foreign trade. Briefly, what characterizes this school of

Table 3.14 *Production and trade in CNC lathes in the FRG and France, 1976*

	FRG	France
Production (US$ m.)	98.1	32.8
Exports (US$ m.)	79.2	11.8
Imports (US$ m.)	10.1	10.8
Apparent consumption[a] (US$ m.)	29.0	31.8
Exports/production (%)	80.7	35.9
Imports/apparent consumption (%)	34.8	33.9

Sources: Elaboration of data supplied by CECIMO and Eurostat, *NIMEXE Analytical Tables* (1978).
[a] Production minus exports plus imports.

thought is the assumption that technology is not a free good but that the creation of proprietary knowledge in the form of some kind of industrial breakthrough can give firms and countries a headstart in a particular line of business. The initial headstart can then be prolonged by the existence of economies of scales) see, e.g., Soete, 1981; Hufbauer, 1966). In Hufbauer's words: 'Technology gap trade is therefore the impermanent commerce which initially arises from the exporting nation's industrial breakthrough, and which is prolonged by static and dynamic scale economies flowing from that breakthrough' (Hufbauer, 1966, p. 29).

In the context of the CNC lathe industry, it is perhaps less meaningful to use the term industrial breakthrough. However, the early Japanese concentration on less sophisticated CNC lathes and their quick exploitation of the new opportunities created by the developments in electronics can be seen as the development of a new product, a new technology. The production of CNC lathes is associated with major economies of scale (which will be discussed in Chapter 4); the Japanese strengthened their position vis-à-vis their overseas competitors by being the first to capture the hitherto only potential benefits of economies of scale. Thus, although the initial headstart given by the development of proprietary knowledge was eroded by competitors copying the Japanese designs, the Japanese firms had gained a very good position in terms of cost competitiveness. Given this situation, the response of the European firms was to some extent to withdraw to a defensive position and try to exploit the market segments in their domestic markets that the Japanese had opened up. This is borne out by the fact that, between 1976 and 1980, the export ratios of the FRG, Italy and the UK dropped significantly. This can be seen in Table 3.15. The exception is France, which increased its export ratio, although it reduced it in the 1980s. The adjustment process in Europe is further discussed in section 3.4.

Table 3.15 *Exports as share of value of production of CNC lathes in four EEC countries, selected years, 1976–1984*

Year	FRG	Country Italy	UK	France
1976	80.7	44.8	69.5	35.9
1980	57.7	31.1	42.1	65.6
1982	60.5	39.7	46.9	26.3
1984	67.2	50.6[a]	33.7	20.0

Sources: 1976 and 1980: elaboration of data supplied by CECIMO and Eurostat, *NIMEXE Analytical Tables* (various).
  1982: NMTBA (1984/85), MTTA (1983), VDW, Unione Costruttori Italiani Macchine Utensili.
  1984: FRG – data from VDW and CECIMO; Italy – Eurostat, *NIMEXE, Analytical Tables* (1983) and data from CECIMO; UK – MTTA (1985); France – MTBA (1985/86).
  [a] 1983.

Thus, the lack of imports of CNC lathes to Japan in the second half of the 1970s could be explained at least partly by the lack of effort by European and American firms to export to Japan. Indeed, several Japanese in the machine tool industry claim that they rarely see any advertisements from foreign builders and that foreign firms are not prepared to take the initial cost involved in penetrating a new market.[5]

(iv) Finance A final factor of interest in discussing the success of some Japanese firms is the possible role of financing in permitting firm behaviour that involves taking high risks. A recent study of the Japanese machine tool industry claims that one major factor behind the success of the Japanese machine tool industry is that: 'Management sets long range goals for market share and output and takes a world wide view of the market and willingly sacrifices short term profits to achieve their long term goals' (*Metalworking, Engineering and Marketing*, 1982, p. 86).

The implementation of a strategy that includes high growth rates may also involve taking high risks. The risk-taking includes not only the R&D phase of individual projects, as interpreted by Watanabe (1983, p. 43), but also getting geared up to and implementing such rapid growth rates as shown in Table 3.10. Implementing such high growth rates normally implies increases in fixed capital and, indeed, several of the Japanese firms moved into new plants at the end of the 1970s. Perhaps more importantly, it implies an improvement of marketing capabilities, which involves a large amount of fixed costs.

One of the findings from the US machine tool builders' study mission to Japan in 1980/81 was that: 'Management is willing to make the up-front investment in terms of marketing costs, establish sales offices, invest in finished product inventory for quick deliveries, provide stock for spare parts for immediate delivery and mount aggressive advertisement campaigns' (*Metalworking, Engineering and Marketing*, 1982, p. 89).

The design of company strategies based on an assumption of future very fast growth rates in the demand for CNC lathes distinguishes Japanese companies sharply from European ones. For example, the marketing manager of a major European CNC lathe builder exclaimed that 'had we known about the market developments in CNC lathes, the Japanese would not have had a chance'. As I argued above, these Japanese firms opened up new markets and at least to some extent created new markets and therefore caused the market to develop as it did. Hence, in contrast to the European builder, the Japanese saw the size of the market as a variable and not as a parameter.

Of course, the perceived risks associated with a given strategy are partly a function of the amount of information the decision makers have. A manager in a firm possessing superior information on, for example, market needs may well implement a strategy that is perceived as too risky by managers in other firms (see Ansoff, 1977). The common assumption in economics of free information is probably one of the least valid ones.

There may, however, be other factors of a financial nature involved in explaining the different behaviour of firms. One study (*Metalworking, Engineering and Marketing*, 1982) suggests that, in comparisons of US and Japanese firms, the tendency of firms in the latter country to make long-term investments is partly a function of their greater reliance on bank loans than on sales of securities to meet their capital requirements. The logic behind such a statement would be that the interest payable on loans is deducted from a firm's income prior to establishing the taxable profit. Dividends on stocks, on the other hand, have to come from the profits left over after the business tax has been paid. Since one of the alternatives for shareholders is to put their money in a bank instead of buying stocks, the pressure for a firm to pay a dividend equal to the going interest rate is high. The cost for firms relying on sales of stock to finance new investment is therefore higher than for firms relying on bank loans.

Comparing, however (in Table 3.16), the largest Swedish and West German firms with some of the largest Japanese firms, we find that the ratio of own capital to total capital did not vary much around 1976–79. Furthermore, if we compare the share of *debts* in total capital in 1979, the two European firms had a *larger* share of debt in total capital than their Japanese counterparts (see Table 3.17). On the other hand, firms with a lower debt ratio would tend to be able to take higher risks than firms with a higher debt ratio. The lower debt ratio of the Japanese firms in 1979 could therefore be one factor behind their greater propensity to take risks.

Table 3.16 *The ratio of external capital to total capital^a in some CNC lathe producers (%)*

Year	SMT Machine Co (Sweden)	Gildemeister (FRG)	Okuma (Japan)	Hitachi Seiki (Japan)	Ikegai (Japan)	Mori Seiki (Japan)
1976	65	n.a.	78	67	90	n.a.
1977	68	n.a.	n.a.	n.a.	n.a.	n.a.
1978	66	87	n.a.	n.a.	n.a.	n.a.
1979	69	87	70	75	90	n.a.
1980	72	83	66	60	88	n.a.
1981	77	81	54	63	89	n.a.
1982	90	n.a.	41	52	89	13
1983	74	n.a.	29	51	96	8
1984	75	n.a.	44	59	−16	30

Sources: SMT Machine Company and Gildemeister – annual reports; Japanese companies – The Oriental Economist (various years).
^{a} 1 − (equity capital/total liabilities + equity capital).

Table 3.17 *Debts as a percentage of total capital^a in some CNC lathe producers*

Year	SMT Machine Co (Sweden)	Gildemeister (FRG)	Okuma (Japan)	Hitachi Seiki (Japan)	Ikegai (Japan)	Mori Seiki (Japan)
1977	n.a.	48	n.a.	n.a.	n.a.	n.a.
1978	48	32	33	n.a.	n.a.	n.a.
1979	48	38	10	30	46	n.a.
1980	49	33	9	17	43	n.a.
1981	51	n.a.	0.6	12	43	0
1982	n.a.	n.a.	0.3	10	47	0
1983	50	n.a.	0.3	19	60	0
1984	53	n.a.	0.1	21	73	0

Sources: As for Table 3.16.
^{a} Debts are totals of long- and short-term borrowings. The external capital is a broader concept as it also includes, for example, Bills payable to component suppliers, etc. The debt share in external capital is significantly higher for SMT Machine Co and Gildemeister than for the Japanese firms, implying that the cost of capital is lower for the Japanese firms.

Another factor thought to be of some importance is that 'Banks are often major stockholders and primarily interested in a company's long term growth. Pressure from stockholders is therefore weak for yearly profit' (*Metalworking, Engineering and Marketing*, 1982, p. 89). Institutional differences could therefore mean, *ceteris paribus*, that a machine tool firm in Japan would have a higher propensity to take risks than firms in the USA. It is true that three of the largest firms in Japan are owned by banks and other financial institutions. However, these banks did not seem to inject capital into the firms during the period 1975–79. At least, there was no change in the firm's own capital in this period. The remaining two large firms are mainly family-owned. Hence, whilst the data would be consistent with the hypothesis that some of the Japanese firms are under less pressure to produce profits than their US counterparts, the data do not support the hypothesis that an agent, external to the machine tool firms absorbed a lot of the risks of implementing a strategy involving high growth rates either through loans or through equity expansion.

Indeed, one of the most expansionary Japanese firms, which is one of the two leading firms in the industry today, claims that it sold a hospital in order to be able to invest in a new plant at the end of the 1970s (interview). Thus, the expansion appears to have been based on internally accumulated funds.

(v) Conclusions Summing up this tentative discussion on the reasons behind the success of some Japanese companies, I have emphasized the role of the domestic market in Japan – in terms both of its early demand for less complex NCMTs and of its closed (in a descriptive sense) character vis-à-vis imports of foreign made NCMTs. A further important factor was the early development of cheap CNC units in Japan, in which the availability of a domestic electronics industry was a prerequisite. However, it appears as if the role of the electronics industry was one of responding to demand rather than one of pushing the reorientation of machine tool design via the developments on the electronics side.

A strategic change is a change in the product/market mix of a firm. As was argued in section 3.2, the process of strategic change is often a protracted and cost-consuming process (Ansoff, 1977). Uncertainties are great and risks can be perceived to be high. The greater the similarity between the present strategy and the new strategy, the closer are the similarities between the resources involved and the smaller is the information gap and, hence, the risks perceived. Firms pursuing an *expansionary change* face smaller risks than firms pursuing a *planned diversification*. The former involves a growth in sales given the same basic strategy and the latter involves a clear change in product/market mix. One could well argue that, given the very early Japanese orientation towards smaller and more simple NCMTs, the change involved in choosing the *overall cost leadership strategy* in the mid-1970s did not entail as great uncertainties as it would have done for the average European firm. The Japanese strategic choice was closer to an expansionary change than to planned diversification. The risks involved in the strategic change were further reduced by the fact that imports of CNC lathes have always been minimal.

Finally, I found that the Japanese firms, contrary to expectations, had a lower debt ratio than two of the larger European firms. I also found that the expansion of the Japanese firms was not based on the sale of new stocks in the period 1975–79. In conjunction with the relatively low debt ratios, we may possibly conclude that the firms financed the expansion and took the associated risks themselves. The fact that banks and other financial institutions are the owners of three of the leading firms could be another factor explaining the attitude to risk-taking.

3.4 The European response

The implementation of the overall cost leadership strategy by some Japanese firms has fundamentally altered the nature of competition in the international CNC lathe industry. One very important aspect has been that price competition has increased in importance as a competitive tool. Firms that do not wish to or cannot follow the overall cost leadership strategy, therefore have to find other strategies for survival or they will have to leave the industry. This is not an altogether uncommon choice and, if the state in several countries had not been so generous in supporting their machine tool industry, the bankruptcies would have been far more frequent.

To understand the scope for response by the existing firms as well as the scope for new entrants, we need to begin by going back to the discussion in section 3.2 about how different market segments demand different types of technology. The basic factor conditioning the scope for response is the segmentation of the market for CNC lathes. Indeed, there is extensive product differentiation within the industry, where each type of CNC lathe serves special submarkets. The demand from these submarkets varies with respect to a number of factors:

- the performance of the lathe
- the size of the lathe
- the degree of standardization of the lathe.

GROWTH AND MARKET STRUCTURE 67

For example, a large firm – say an automobile firm in the FRG – demands a custom-designed CNC lathe of a very high performance equipped with an automatic material handling unit. Alternatively, the customer may be a general machinery firm, such as a pump manufacturer, which demands a medium-performance, smaller CNC lathe. Finally, the customer may be a jobshop that demands a very versatile, large CNC lathe of a standard nature that is simple to use.

The fact that the market is rather segmented means that a firm can avoid, or at least lessen, the competitive pressure from the Japanese by exploiting some particular market niche. In particular, price competition can to some extent be avoided by serving market segments with a lower price elasticity of demand than those served by the Japanese firms.

3.4.1 Overall cost leadership strategy

One Italian firm has opted to follow the overall cost leadership strategy, at least partially. In 1981–82 this firm produced a series of three models of CNC lathes with a low performance (in the sense of motor power but not in precision) – indeed, lower than the average Japanese CNC lathe – and marketed itself as the *European challenger*. In the early 1980s the volume of output reached over 400 units, which were primarily sold to the local market. In 1982, the firm was so indebted to Olivetti, which makes CNC units, that Olivetti took over the firm. It can now be characterized as a follower of the overall cost leadership strategy but its characteristics are not identical to the Japanese. The firm produces 350–370 units per year of basically standard CNC lathes. It also sells a few FMMs, but such sales came to less than 10 units in 1984. Since 1982 it has extended its product range to larger models with greater horse power, but the bulk of production is still accounted for by smaller models with fairly weak motors. Around 80 per cent of its market is constituted by jobshops, although it also sells to larger firms such as Fiat and General Motors. The firm dominates the Italian market and estimates that it accounts for around 40 per cent of sales to the Italian market. This market is around 500 units per year and is thus rather small, so the firm has increased its export share from 28 per cent in 1982 (in units) to 48 per cent in 1984.

In terms of its product/market choice it is very similar to the leading Japanese firms. The volume of output is, however, much smaller. In spite of this the firm has been profitable since 1983. This is probably to a large extent due to the advantages of being owned by Olivetti. Olivetti is a producer of CNC units and other machine tools, e.g. machining centres: in 1984 it made 4,000 CNC units for lathes as well as 220 machining centres. The lathe producer always gets controls cheaper than do other firms buying Olivetti CNC units. For smaller

CNC lathes, the cost of the control unit accounts for a large share of the production cost; hence, access to cheap CNC units must affect the firm's profitability in a positive way. Furthermore, the firm can draw upon Olivetti's knowledge in software development, which is probably the main reason why they can manage with less than twenty design engineers. Finally, the production of the lathes is done in a factory that also produces machining centres, which means that production can be rationalized somewhat.

An interesting aspect is that the Japanese share of the Italian market is much smaller than its share in other countries (see Table 3.18). Although there may be several reasons for this, one is probably the similarity between this firm's products and the Japanese products.

Table 3.18 *Share of all imports and those of Japanese origin in the apparent consumption of CNC lathes in some OECD countries, 1980 and 1984 (% of value)*

Country	Share of all imports		Share of imports of Japanese origin	
	1980	1984	1980	1984
France	69.2	48.1[a]	16.3	7.6
UK	66.7	70.4	30.5	28.4
Sweden	60.5	55.9	32.5	27.5
FRG	37.5	30.0	19.8	13.6
USA	29.0	49.5	23.0	42.9
Italy	18.3	27.4[a]	1.8	3.9[a]

Source: Elaboration of national and regional production and trade statistics.
[a] 1983.

3.4.2 Focus strategy

A different strategy pursued by a small number of firms involves focusing on the requirements of other segments of the market. Some West German firms have concentrated on the demand for high-performance CNC lathes from what is still a substantial part of industry, the part that demands a high cutting capability and very high precision. High-performance CNC lathes are required not only by some larger firms but also by medium-sized and even small firms. Among the latter firms are, for example, subcontractors to automobile companies, which, even though they may be small, demand a different type of technology than a similar sized firm making a range of different products, i.e. a job shop, which may instead demand an easy-to-use, very versatile and not necessarily very high-powered type of CNC lathe.

Firms pursuing the high-performance focus strategy are, however, by no means isolated from price competition as there is a degree of substitutability between high- and medium-performance CNC lathes, the latter being produced by firms pursuing the overall cost leadership strategy. Firms following the high-performance focus strategy have therefore sometimes adjusted performance downwards somewhat – introducing smaller models and standardizing their products to a greater extent than before. This is illustrated by a quotation (freely translated from the German) from the annual report of one such firm:

> The programme of the firm includes the area of high performance CNC lathes of the series MD and the smaller standardized models of the series MD 5S and MD 10S. Through the introduction of the new series MD S, with inhouse developed CNC units of modern modular building blocks, the market position of the high performance lathes can be extended. Especially positive was the ability to benefit from many advantages of production in larger series allowed by the use of principles of standardization. This resulted in an advantageous price/performance ratio in the market. The model MD 5S became, in 1979, our most important product. [Gildemeister Aktiengesellschaft, 1979, p. 16]

Standardization has been combined with the option of some custom design by using modular design. The firms have also been able to increase their sales volume and have therefore reaped at least some benefits of scale economies. For example, two German firms have output levels of 500–600 units per annum. Another firm, which operates as a division within a larger machine tool firm, produces 250–300 units per annum. Let us look more closely at two German firms following this strategy.

The first firm has around 2,000 employees and is owned by a family. Its products are called the Mercedes Benz of CNC lathes. Its market is thus the more demanding one in terms of performance and precision. Its machines are based on modules and therefore vary from each other. It has always tried to standardize its CNC lathes and has found a good compromise between standardization and custom design through using modules. It produces 500–600 CNC lathes per year, which is a dramatic change since the end of the 1970s, when only 20 per cent of its output was CNC; today it is 80 per cent. It also produces some automatic, conventional lathes. A lot of developments have taken place in recent years as regards attachment and peripheral equipment. For example, the firm has developed an internal gauging system as well as rotating tools. Another important feature of its development is FMMs. Although 50 per cent of its output consists of CNC lathes that do not use robots for material handling (bar machines), around 50 machines annually (of 250) are sold with automatic material handling equipment.

Although the markets it serves include almost all large companies, the majority of its machines are supplied to smaller and medium-sized firms.

As far as the CNC unit is concerned, the firm uses Siemens hardware and its own software for single-spindle CNC lathes. For the multi-spindle CNC lathes, the system is entirely the firm's own. The advantage to it of designing its own software for the units controlling the single-spindle machines is that it can incorporate functions in its CNC lathes that would not have been possible had it relied on standard CNC units. The reason for developing its own control unit for the multi-spindle machines was essentially the same – it was not available in the market.

The firm has a hundred design engineers, plus thirty-five to forty software people who also do system development.

The firm has been profitable for thirty-five years. One reason for this, it is claimed, is that it does not sell FMMs to distant firms. Because FMMs are to a considerable extent custom designed and still involve debugging problems, they are very engineering-intensive. Naturally, the cost of such engineering work increases if sales are undertaken at a distance from the factory, unless sales in a particular market are large enough to justify a separate organization in that market.

The second firm has 1,300 employees in the FRG. It also has nine subsidiaries abroad. Production of CNC lathes began in 1979. Sales from Germany amount today to around 200 million DM or roughly 70 million USD. It produces standard CNC lathes and automatics as well as conventional automatic lathes and also offers a wide range of toolings and accessories. Its machines are based on modules. Annual output is 500–600 CNC lathes and around 300 automatic lathes. Of the CNC lathes, 50–70 are equipped with flexible handling systems.

In terms of the CNC unit, the firm approached different CNC-controlled manufacturers to make a custom-designed unit but they could not comply with its demands. So the firm made an agreement with Mitsubishi, which now produces a special CNC unit for them. The software is, however, designed by the German firm. The advantages of having an inhouse-made unit are that

- there is only one source of service for the CNC lathe,
- the firm can develop special features such as automatic tool compensation, where it now feels that it is ahead of its competitors.

It has fifty to sixty design engineers, of whom twenty to twenty-five are developing software. In addition, it has ten to fifteen who develop systems. The firm claims to break even and indeed that it is in a very prosperous position.

It is of interest to note that the European firms that follow the overall cost leadership strategy partially and the focus strategy operate in contexts characterized by (a) a reasonably sized domestic market (in the case of West German firms only; see Table 3.12) and (b) a relatively low level of imports (see Table 3.18, which shows that German and Italian imports are considerably lower than those of the UK, France and Sweden). I noted in section 3.3.1(iii) that one important factor determining the success of the Japanese firms has been the lack of imports of CNC lathes combined with a large domestic market. The same feature of a large domestic market (for West Germany only) which is fairly well catered for by local suppliers would seem important in inducing firms to pursue these two strategies. Of course, with domestic firms pursuing these strategies, imports can be kept low. There may also be other factors at work determining the import ratio, such as state policy. Jones (1983) describes, for example, the Sabatini Law in Italy which acts as a subtle protection for Italian-made machine tools: certain financial advantages are received only in association with the acquisition of Italian-made machine tools.

3.4.3 *Differentiation strategy*

Firms in smaller countries, e.g. Sweden, or in countries with an historically given high import share, e.g. Britain, where the share of imports in demand was already 76 per cent in 1975, have often opted for a different adjustment policy. One Swedish firm, for example, has gone in for the pure custom-design strategy, producing fewer than 50 units per year. The market for such products is, however, so small that firms that follow this strategy account for only a marginal share of the output of CNC lathes.

A more common strategy is to *differentiate* the firm's output from that of its competitors through continuing to emphasize product development as the main source of competitive strength. (Note that this was a characteristic of firms in the early phases of the product cycle.) Indeed, one Swedish firm's strategy involves changing the product, at least partly, by developing FMMs, i.e. a system development where a unique material handling unit is attached to the CNC lathe thereby enabling the CNC lathe to be run unmanned for up to six hours. The unit is marketed with high-performance CNC lathes. In the initial stages of production of these modules, production volumes can be fairly small because performance, not price, is most important for the customers. The types of customer demanding this technology are again the *leading edge* firms discussed in section 3.2. Let us look a bit closer at this firm as well as at a UK firm that follows a similar strategy.

(i) A Swedish firm The firm is the largest Swedish firm producing CNC lathes: in 1984, it had 431 employees and sales of approximately 30 million USD, of which 24 million derived from sales of CNC lathes. It has been a part of a large state-owned group of machine tool builders since 1970.

The first NCMT produced by the firm was a milling machine in 1957 and by 1965 a rudimentary capability had been achieved in the production of numerical control units. The firm entered very early into the production of its own CNC units with the help of a computer manufacturer that saw the machine tool market as a good potential market. In a reconstruction after a crisis in 1970, it was decided to put great emphasis on CNC lathes based on a minicomputer. Break-even was planned for 1975 but was achieved by 1973. By 1974, the firm produced only CNC lathes and recorded 18 per cent of its sales as profit.

By 1975, the firm recognized that it had a technological lead: 'The firm has several hundred machine systems in use while competitors are presenting their prototypes' (SMT Machine Company, 1975, p. 2). This extremely important point was, however, made *en passant* in the Annual Report and the tone of the reports at this time was one of pessimism for the future. This pessimism kept appearing all through the decade in spite of generally rising sales and the technological lead.

In 1975 and 1976, sums corresponding to 4.5 and 5.0 per cent respectively of sales were taken out of the firm to be invested in other machine tool firms within the conglomerate. Indeed, a firm representative claims that it was generally so that profits were taken out of the firm in the 1970s to subsidize other activities in the conglomerate. Investment was low in the years after 1976.

By 1979, the Japanese were mentioned in the Annual Report for the first time as a big threat, although sales of the firm's CNC lathes kept on increasing. In the same year, a change in the R&D direction was implemented and the majority of the R&D resources were allocated to develop the material handling unit mentioned above, the Computerized Parts Changer (CPC). The firm thus began to develop a Flexible Manufacturing Module (FMM). This was at the same time as the German firms, discussed above, reacted to the Japanese threat by standardizing and sometimes reducing the performance of some of their CNC lathes, but above all by expanding their production greatly. The focus on the CPC was further strengthened in 1980 when a large investment was made to increase the assembly space for the CPC.

In the course of the last few years, the firm has achieved annual sales of 50 FMMs, which constitutes roughly one-third of its sales in units; 50–55 per cent of the FMMs are exported. The firm hopes to be able to reach annual sales of around 100 FMMs within a few years.

The firm thus, had an early technological lead, which was strengthened by the high level of demand in the home market, where the diffusion of CNC lathes was faster than in other countries (as was shown in Table 2.5). The market was, however, strongly dominated by larger firms, smaller firms having only 3 per cent of the stock of CNC machine tools in 1976 (Steen, 1976). The firm failed, however, to exploit these advantages and create a long-term competitive advantage based on reaping the benefits of scale economies. In spite of its very early specialization in CNC lathes, the firm produced only around 250 units per year in the early 1980s, a figure that now has dropped to less than 200. By 1981, price competition from the Japanese and others had become very notable and profits dropped to only 3.5 per cent of sales. By 1982, the firm began to make people redundant as profits became negative. Significant losses were made in the years 1982–84.

An important factor behind this failure was the continuing pessimism concerning the market potential for CNC lathes. This pessimism is very noticeable in nearly all Annual Reports in the 1970s, in spite of it being proved wrong nearly every year. This is in contrast with a leading Japanese firm that had very large losses in this period. Despite these losses, the firm decided to specialize in CNC lathes and in 1977 it made a large investment in both its production and marketing capacity. Upon being asked how it could take such a risk, it claimed that it knew that CNC lathes would be a great success. Of course, its success could be based purely on luck, but it is more probable that it had a better understanding of the changing market than did the Swedish firm.

Associated with this basic view of the market potential of CNC lathes was the transfer of funds from the firm in 1975 and 1976 and the low investment in subsequent years – precisely the years when the Japanese were preparing for their expansion. The manager of the Swedish firm at that time claims that the firm had wanted to expand more in those years but that it was not given the investment funds (interview). Hence, when the expansion of the CNC lathe market really took off in 1976–77, the firm rapidly lost market shares, as is shown in Table 3.19. (The table shows production data for all of Sweden, but the firm heavily dominates Swedish production.)

The firm has always designed high-performance CNC lathes, as was common in Europe. However, while several German firms adjusted their output mix to include CNC lathes of more standard type produced in larger volumes by 1978–79, this firm decided at this time to allocate most of its R&D resources, about fifty to fifty-five people, to developing the CPC unit. Hence, whilst the trend in the world market was to focus on the design requirements of the smaller and medium firms with a higher price elasticity of demand, the firm again took demand from the *leading edge* firms as its point of departure in its design work. In conjunction with a low volume of output, this has resulted in poor competitiveness in the standardized high-performance segment, as a result of being threatened both by cheap medium-performance Japanese CNC lathes and by West German builders that have adapted their strategies. Today, the firm sees the system approach (where it is making money) as the solution to its problems in the standardized market segments (where it is losing money) and is trying in this way to avoid price competition.

Table 3.19 *The international specialization and world market share of the Swedish CNC lathe industry, 1974–1984*

Year	International specialization[a]	World market share[b] Value %	Units %
1974	−0.03	—	—
1975	−0.23	4.63	6.91
1976	0.17	4.65	6.22
1977	0.13	4.00	4.67
1978	0.25	3.25	3.06
1979	0.31	3.48	2.93
1980	0.16	2.49	1.58
1981	0.02	2.63	1.56
1982	−0.16	1.56	1.20
1983	−0.20	1.48	1.12
1984	−0.13	1.39	0.87[c]

Source: Swedish trade statistics were used to calculate the international specialization. The world market share was calculated on the basis of data supplied by the Swedish association for the machine tool industry as well as data from CECIMO.

[a] (Export − import)/(export + import).
[b] World market share refers to the share of Swedish production in the production in Japan, USA, the FRG, France, Italy, the UK and Sweden.
[c] Assumed production of 200 units.

This historical example illustrates two aspects of strategic change. First, it demonstrates the importance of continuous strategic revision. The decision to focus on the CPC in 1979 was in all likelihood not based on a perception of a competitive threat from the Japanese; the firm prospered in that year, with over 11 per cent profit on sales. Rather, the choice of a system approach should be seen as a continuation of a strategy where product development was emphasized as the strength of the firm because the firm was probably unaware of how the nature of competition changes through the product cycle. The choice of CPC should therefore not be seen as a long-term strategic change. Today, the firm has been overtaken by events, and clearly identifies system development as its strategy.

Secondly, the example illustrates the problems for firms in smaller countries that follow in the wake of the maturation of a product. Even though the specialization ratio became positive for Sweden in the mid-1970s, largely owing to this firm, the Swedish world market share has constantly declined (see Table 3.19). Furthermore, after 1979, when the firm began its reorientation towards the system approach, the positive specialization weakened considerably and turned negative from 1982. The market share also dropped dramatically.

Of particular importance is that, if the firm does not already have a marketing network, an expansion based on foreign markets involves not only great costs but also greater uncertainties, and therefore risks, than an expansion of sales in a known domestic market (Hirsch, 1967). As is discussed in more detail in Chapter 4, there are important barriers to entry to overcome as a new market is penetrated. This means that a firm in a small economy will always need to take greater risks or possess better information than firms with a larger domestic market.

Thus, at critical times in the evolution of the industry, such as in the mid-1970s for this firm, a considerable amount of risk capital would have been necessary in order to provide the firm with an additional home market, e.g. the US market. In 1980, the Swedish firm produced around 250 units per annum and exported about two-thirds of its output (i.e. about 170 units). If the firm were to increase its production to say 500–600 units per year, which would probably have been necessary in order for it to stay competitive in the standardized high-performance segment, it would have needed to increase exports by 250–350 units. Such an expansion would have required considerable investments in additional marketing and production facilities and skills as well as a management who saw the need for an expansion and could find the funds for it. Instead, as I mentioned above, resources were taken out of the firm in the 1970s!

Today the firm has a small technological lead again. However, its competitors, both large Japanese and German firms, are developing similar systems. At present, the cost structure of the FMMs (suggested by the Boston Consulting Group, 1985, p. 51) typically includes more than 60 per cent hardware and less than 40 per cent software. A German firm similarly suggests that the software amounts to about one-third of the cost of the system, with another third being accounted for by the CNC lathe and the final third by the material handling equipment and other peripherals (interview). The software is custom designed although basic modules can be used. The development of such modules is to some extent based on the past sales experience of the firms, and the leading firms in this business may have a competitive edge in that they have progressed further down the learning curve than their competitors.

Although the systems are custom designed, the bulk of the costs are still accounted for by more or less standard items. Estimates vary, but the Swedish firm suggests that around 50 per cent of costs are accounted for by the CNC lathe and parts of the material handling unit, which can be produced for stock; only one element of the material handling unit needs to be custom designed. Another 35 per cent are accounted for by ancillary equipment, which is often bought in; these items vary slightly between customers. Most of the remaining 15 per cent, which is software, is now standardized for this firm, although we should note that this may not be the case for their competitors as yet. Thus, for this firm, the bulk of the costs for FMMs are already comprised of standard items where economies of scale can be reaped. As its competitors' systems become more standardized, the overall cost competitiveness of these firms, deriving from their larger volume of production, will affect the cost of the systems too. Hence, the Swedish firm needs to get volume production of both the CPC and the CNC lathes.

As we saw above, its German competitors are rapidly accumulating experience in the FMM field and so in all likelihood are some Japanese firms. Indeed, the Boston Consulting Group argues in a similar way in relation to European industry:

> Japanese suppliers currently have some cost advantages in the basic machine tools . . . On the assumption that Japanese builders also have access to cheaper electronic hardware and a similar cost position in software and tooling, there is a considerable production cost advantage of Japanese over European suppliers in flexible cells. If they are to play a major role in the business, EEC suppliers will have to:
> — improve their competitiveness in standard machine tools and material handling devices;
> — become a leading force in standardizing software for flexible machine cells;
> — exploit opportunities for competitive advantage in tooling and other peripheral equipment. [The Boston Consulting Group, 1985, p. 5]

However, the Swedish firm not only needs to develop volume production of standard CNC lathes *and* further develop its system sales, it probably also needs to strengthen its design team. With the recent growth in the size of the leading firms in the industry, this firm may encounter problems in keeping up in the technological race. The firm has around fifty-five designers, including CNC designers, as compared to two to three times this many in some Japanese and German firms. Thus, the competitive position of the firm dictates a large injection of capital in order to exploit its new technological lead.

The management is well aware of the risks associated with its present strategy, but a strategic reorientation of the kind outlined above would require more capital than the management considers it feasible to acquire. The management hopes instead that it will be able to stay a product developer in the future through identifying even more advanced systems.

(ii) A UK firm The UK firm has around 450 employees and estimated sales of about £15 million in 1984. It produced 190 CNC lathes in 1984. The firm therefore accounted for 50 per cent and 23 per cent of UK production in terms of value and units respectively. It is a specialized machine tool firm but is owned by a larger engineering group. The firm began producing CNC lathes at the end of the 1960s and its main market has been in the automotive and energy industry, which demands large, powerful and often custom-designed CNC lathes. The main product in the early and mid-1970s was a CNC lathe for the automotive industry. After 'the Japanese showed the way', as a representative put it, the firm developed a standard CNC lathe in 1978. This lathe was considerably smaller than its previous ones, although still not as small as many of the Japanese CNC lathes; it was also a high-performance lathe, or rather a series of three models. In 1982, the firm developed a medium-performance, smaller CNC lathe. At present, the custom-designed, very large CNC lathes account for around 20 per cent of the units produced, although considerably more in terms of value, and the remaining 80 per cent of the output consists of standardized CNC lathes.

The strategy of the firm rests on its strength in application engineering – in designing systems as well as special equipment for handling work pieces, tool holding and ancillary equipment. Its main market is therefore larger firms. Special engineering is made on both the larger and the smaller CNC lathes, and they account for 50 per cent of the sales in terms of value. It is of interest that the firm supplies a fair number of CNC lathes with automatic material handling units attached. Since 1984, 25 such lathes have been sold, of which 15 were its own make and 10 were CNC lathes of other makers that were incorporated into larger systems that it supplied. The firm uses a standard CNC unit because its application engineering task does not involve designing completely new production processes, but rather consists in putting together existing technologies in a new way. The firm does, however, produce the software that connects the material handling robot to the lathe.

The management is aware that a strategy based on pure application engineering will not suffice in the medium term. As the standardization of the ancillary equipment and software proceeds, the price of the central component in the system, the CNC lathe, will become the key to gaining orders. Indeed, the firm claims that, already in 1982, price competition had begun to be apparent in the sales of systems too (for example, four standard CNC lathes linked by ancillary equipment). Furthermore, sales of a good range of standard products are seen as necessary in order to be able to supply the engineering capability required for system sales and development. Hence, whilst the strength of the firm rests on system development, it also realizes that it needs to expand the production of standard CNC lathes.

The firm is a good example of the problems involved in adjustment. The firm traditionally chose to build CNC lathes with a very high unit value. This appears to be the case generally with UK firms. Table 3.20 shows that, while the world market share of the UK in terms of value stayed roughly constant between 1975 and 1981, the share in terms of units was halved. After 1981, the UK's share of world output in terms of units rose a lot, but that can be explained by a rise in the production of very cheap teaching CNC lathes with a very low unit value. Indeed,

Table 3.20 *The international specialization and world market share of the UK CNC lathe industry, 1975–1984*

Year	International specialization[a]	World market share[b] Value %	Units %
1975	−0.19	3.75	4.03
1976	−0.17	3.55	3.66
1977	−0.36	3.19	2.30
1978	−0.43	4.17	2.87
1979	−0.61	3.80	2.43
1980	−0.47	3.97	2.13
1981	−0.41	3.21	2.15
1982	−0.35	4.84	3.56
1983	−0.37	3.12	4.35
1984	−0.64	2.65	3.56

Source: Data supplied by the British Machine Tool Trades Association and CECIMO.
[a] (Export − import)/(export + import).
[b] The *world* market is defined as the output of the seven countries listed in Table 3.1.

one firm, which is the leading firm, sold 1,000 units in the period 1982–October 1985 (interview).

The differentiation strategy based on systems is probably not a viable strategy in the medium term as, over time, price competition will become stronger. As in the case of the Swedish firm, this firm needs to go for a higher volume of production *as well* as for system development. In the Swedish case, the risks involved in an adjustment were increased by the small size of the domestic market. In the UK case, the home market is larger and fast growing – from 222 units in 1976 to 1,449 units in 1984, although a fair share of these are simple teaching machines. The share of imports in the total investment in CNC lathes has however traditionally been very large in the UK – 79 per cent in 1980 and 59 per cent in 1984 in terms of units (in terms of value the corresponding figures were 68 and 70 per cent). The deteriorating international specialization ratio in combination with a constant world market share (shown in Table 3.20) also indicates the slowness of UK producers to adjust to growing domestic demand. Hence, the benefits of having a large domestic market can only be reaped by firms that stay technically advanced and with products that are in line with market requirements.

3.5 A note on the US producers

This study is mainly based on firm interviews in Europe, Japan and in three NICs. Secondary material has been used extensively where it is available but without detailed information on activities at firm level it is impossible to make an analysis of this kind. For financial reasons, it has not been possible to interview US firms. However, from bits and pieces of secondary material, I can nevertheless say a few words about these firms.

It was shown in Tables 3.1 and 3.2 that the US CNC lathe industry has lost most from the structural change in the global industry over the last few years. Its share of world production dropped from 36.2 per cent in 1975 to 6.7 per cent in 1984 in terms of units produced and from 47.8 per cent in 1975 to 14.0 per cent in 1984 in terms of value.

The import content of investment in CNC lathes is also significant. In terms of units, the import share of investment was as high as 49 per cent already in 1980 whilst in terms of value it was 29 per cent. By 1984, these figures had risen to 70 per cent and 49 per cent respectively (NMTBA, 1985/86). The average price per unit imported was only 70,000 USD in 1980 whilst, in the same year, the average price of CNC lathes produced in the US was as high as 175,000 USD. In 1983, the average price of imported CNC lathes had risen to 62,000 and the average price of US-made CNC lathes was 159,000 (NMTBA, 1984/85). Hence, the large import of CNC lathes into the USA consists predominantly of cheaper CNC lathes, which are mainly supplied by Japanese firms. In 1984, Japan had 43 per cent of the US market in terms of value and around 60 per cent in terms of units. It seems clear that the US builders have permitted a large inflow of CNC lathes that are smaller and of lower performance than those that they themselves produce.

The traditional market for these lathes is the jobshop segment and the medium-sized firms, which traditionally has not been well served by US companies. The president of the world's largest machine tool firm, Cincinnati Milacron, admits that: 'The segment wasn't as apparent to all of us as it was to them' (*Financial Times*, 7 April 1983).

In a situation where there is a sharp rise in demand, as was the case at the end of the 1970s (see Table 3.21), a weak reaction to an increase in imports is not necessarily irrational if short-term profits are sought. The pre-tax net income of machine tool firms as a percentage of sales in the USA rose sharply from around 7.5 per cent in 1977 and 1978 to 12–13 per cent in 1979–81 (NMTBA, 1985/86). It dropped to 5 per cent in 1982 and decreased to nearly a 10 per cent loss in 1983 and a 3 per cent loss in 1984 (NMTBA, 1985/86). The 1979–81 figures are high for the machine tool industry. For the period 1969–1976, the unweighted average profit ratio was only 4.6 per cent (NMTBA, 1984/85).

Table 3.21 *Shipments and unfilled orders of metalcutting machine tools in the USA (US$ m.)*

Year	Shipments	Unfilled orders
1975	1,879	1,062
1976	1,482	1,243
1977	1,651	1,794
1978	2,189	2,981
1979	2,930	4,546
1980	3,681	4,750
1981	4,105	2,873
1982	2,895	1,043
1983	1,372	823
1984	1,607	1,132

Source: NMTBA (1985/86).

Evidence of the poor adjustment by US firms may also be seen from the number of CNC lathes produced by the leading firms. One source

(OECD, 1983) suggests the following output of CNC lathes for the three leading firms in the USA in 1980:

> Warner Swazy 520 units
> Leblond   360 units
> Cincinnati   355 units

Since then the yearly volume of output of all US producers has dropped from 2,739 units to 1,203 units in 1983. Obviously, the annual volume of output of these firms has also had to decrease.

Thus, the evidence, albeit somewhat limited, suggests that the leading US firms are pursuing a strategy where large firms are the main customers and production is concentrated on high-performance CNC lathes produced in fairly limited quantities. Indeed, Sciberras and Payne claim, in the case of US machine tool firms, that '. . . for some leading firms almost half their sales had gone to automobile industry customers in the past. In times of major re-tooling, aerospace and automobile industry customers accounted for the major shares of all the leading US firms' production, often as high as two-thirds of total sales' (Sciberras and Payne, 1985, p. 51). They conclude: 'In general, however, the US industry has not succeeded in significantly reorienting business away from its major traditional customers' (Sciberras and Payne, 1985, p. 51).

3.6 Concluding remarks on the strategies pursued by firms based in the OECD

To summarize this section on firm strategies, I present in Figure 3.3 a strategic map of the CNC lathe industry. On the horizontal axis I have plotted the annual output of eleven producers in Japan and in Europe. The European firms covered around 60 per cent of Europe's output (in units) in 1984. Data from the Japanese suppliers come from 1981 and in that year the share of these five firms in total Japanese production was 66 per cent. On the vertical axis, I have plotted the performance of the lathes produced by these eleven firms. The indicator of performance was motor power.[6]

The overall cost leadership strategy is represented by seven firms – five Japanese and two European. Six of these firms produce more than 800 units per annum and are in the low–medium performance range. I have also included the Italian firm described above in this strategic group. The second European firm in this strategic group is a West German firm that consists of three divisions. Two of these follow

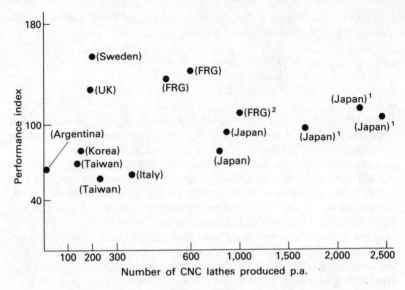

Figure 3.3 *A strategic map of the CNC lathe industry*

differentiation or focus strategies, while the third produces simple lathes. The performance index is an average of the firm's models. The firm does not follow a proper overall cost leadership strategy, but I have placed it among other firms in this strategic group on account of its large production volume. The focus strategy is represented by two German firms, which produce around 500 and 600 units per year. These were also described above. The differentiation strategy, which involves a greater element of product development and custom design, is represented by the UK and Swedish firms described above. As mentioned above, these produce nearly 200 units each.

A third dimension of the figure would include the degree of standardization of the output. It is sufficient to say though that the UK and Swedish firms have a greater emphasis on custom design in their strategy than do the German firms, while the Japanese firms produce largely standardized products. It can also be added that, in the past decade, the leading Japanese firm went from a production of around 300 units per year to over 2,000 units per year. The Japanese industry has as a whole shifted dramatically to the east of the figure whilst the Europeans have moved less dramatically to the east. The occasional firm has moved to the southeast, combining increasing volumes with a slight downward adjustment of the performance of the CNC lathes. As a result of this adjustment by the European firms, the concentration ratio in Europe is estimated to have reached about the same level as that of Japan; i.e. approximately two-thirds of output (in terms of units) is accounted for by the five largest firms.

As far as the US producers are concerned, the limited evidence I have would suggest that they too can be placed in the northwest corner of the strategic map.

While Figure 3.3 gives a broad picture of the strategic position of the leading firms, the main characteristics of the strategies pursued by these firms are summarized in Table 3.22. The characteristics are the same as the decision variables discussed in section 3.2. The ranking of the barriers to entry in each group for a *new entrant* is indicated in Table 3.23. The table includes only a brief statement of the main barriers to entry and I have assumed that the new entrant is one of the eight NIC-based firms discussed in Chapters 5–8. A more detailed analysis of the barriers to entry is presented in Chapter 4.

The *overall cost leadership strategy* involves producing a product of low–medium performance and selling this to mainly medium and small firms with a high price elasticity of demand. Price is therefore low. The marketing is frequently done through independent dealers and the R&D aims at simplification and easy use. Emphasis is also given to designing a product that can easily be manufactured at a low cost, i.e. through a reduction in the number of components in the machine. A large volume of output is required. A few firms produce their own CNC units. The main barriers to entry are economies of scale, access to a large marketing network, and design skills.

The *focus* strategy involves producing high-performance CNC lathes, sometimes with elements of custom design in the form of a choice between various combinations of standard modules. The market is small–large firms with a medium–low price elasticity of demand. Price is medium–high. The marketing is both direct to the end user and through independent dealers. R&D is naturally focused on designing high-performance CNC lathes in combination with standardization efforts (modular design). Sometimes, special application software is developed for particular market segments. System development has begun in the past few years. A medium volume of output is required. Often, the CNC units are produced inhouse. The main barriers to entry lie in design skills, brand image and economies of scale.

The *differentiation* strategy involves either producing CNC lathes or systems with large elements of custom design or providing partially new products such as a new system design. The target groups are the leading edge firms. Prices are high and the volume of output is low. Sales are direct to the user, and communication between the user and

Table 3.22 Summary of the main characteristics of the three strategies[a] pursued by OECD-based CNC lathe producers

Characteristic	Overall cost leadership	Focus	Differentiation
Product:			
Degree of standardization	Standard	Largely standard	Important elements of custom design
Performance	Low–medium	Medium–high	High
Price	Low	Medium–high	High
Target groups	High price elastic	Medium price elastic	Leading edge firms
	Small–medium-sized firms	Small–large firms	
Marketing	Through independent dealers	Through independent dealers or direct to the customer	Direct to customer
R&D	Low-cost, easy-to-use product	High performance coupled with standardization and modular design. Some special designs (software and hardware) developed. Often CNC development. In the order of 10% of sales (in units) are FMMs	Complex, system development sometimes including CNC development. Some special application areas developed. Up to 30% of sales (in units) are FMMs
	Occasional CNC development		
Volume	High	Medium	Low

[a] The names of the strategies are borrowed from Porter (1980).

Table 3.23 *Ranking of the barriers to entry involved in the three different strategies pursued by OECD-based firms producing CNC lathes*

Barriers to entry	Overall cost leadership	Focus	Differentiation
Economies of scale	1	3	
Access to large marketing network	2		
Design skills including electronic design skills	3	1	3
Direct links with leading edge firms			1
Brand image		2	2

the producer is a key factor in the innovative process. R&D is directed towards system development. The CNC unit is sometimes produced inhouse. The main barriers to entry lie in direct links with the leading edge firms, associated brand image and design skills.

Several of the European firms following the focus or differentiation strategies have integrated backwards into the production of the CNC unit. The reason for this vertical integration lies in the importance of electronic design skills in the process of innovation if the firm is extending the frontier. The Swedish firm, for example, claims that it would never have been able to innovate in the way it has without the inhouse production of the CNC unit. Similarly, as we saw above in the case of the German firms, inhouse production of the software enables a firm to incorporate features into its products that otherwise it would not have been possible to do. Furthermore, inhouse production can prove to be cost efficient. Interestingly enough the Swedish firm, with an annual production of less than 200 units, claims that its unit cost is half the price asked by three leading non-Japanese suppliers of CNC units – that is, the price not for standard CNC units but for special ones equalling the performance of the inhouse produced one, which controls a whole FMM. The CNC unit is therefore different from the standard ones used for standard CNC lathes produced for mass markets.

Two Japanese firms also produce their own CNC units. One has been making them for decades and it has used its electronic skills to develop an FMM with a function similar to that of the Swedish firm's. The other Japanese firm claims that it began producing CNC units when Fanuc refused to collaborate in designing a CNC unit that would be extremely easy to use. The availability of such a CNC unit is claimed to be an important element in penetrating the smaller firm market, which cannot afford to have a separate programming office. There are thus various reasons behind the backward integration of firms in this market.

Forward integration into the production of CNC lathes appears to have happened in one case only – in 1981 Olivetti, the large Italian producer of CNC equipment, acquired Pontiggia PPL, the largest Italian producer of CNC lathes. Olivetti is also, however, a producer of machining centres and robots and the acquisition of Pontiggia is seen as a way of being a full-line supplier of factory automation rather than as integration for the reasons mentioned above (*Financial Times*, 27 September, 1983). For the lathe producer, the take-over was however beneficial as it gets preferential access to Olivetti's software knowledge and lower prices on the CNC units.

Having analysed the three main strategic groups within the OECD-based producers of CNC lathes, I shall in Chapter 5 proceed to analyse the strategic position of the NIC-based firms. These firms comprise a fourth strategic group in the world industry. First, however, I shall analyse in more detail, the barriers to entry into the overall cost leadership strategy.

Notes

1. Watanabe (1983, p. 7) was told in interviews in Japan that the introduction of the microprocessor had cut the price of the CNC unit by almost two-thirds.
2. Note that Fanuc developed another NC milling machine with another machine tool firm directly after the completion of the Makino machine (*Today's Machine Tool Industry*, 1976, p. 53).
3. For a discussion of the role of external economies in the early phase of the product cycle, see Hirsch (1967).
4. The price difference between a CNC unit sold by Siemens and one sold directly by Fanuc is claimed to be as high as 35 per cent. See also *Financial Times*, 19 January 1982.
5. This is supported by the reaction of one marketing manager of a larger European CNC lathe producing firm when asked (in an interview) why it did not export to Japan; he looked absolutely horrified.
6. The method of calculating the performance index was as follows. First I assembled technical information on the models marketed by the eleven OECD-based firms and the eight NIC-based firms listed in Table 5.4. The number of models produced by each OECD-based firm is much larger than for the NIC-based firms. I then divided the models into five size groups. For each group, I calculated the average motor power of the group and thereby established an index for each group. As each firm often has CNC lathes in several size groups I proceeded to calculate a weighted average of their normalized performance. To illustrate: a firm has four models – two are in size group 1 and have an average of 1.13 as their performance index; a third

model in size group 2 has an index of 1.05, while a fourth model in size group 4 has an index of 1.0. The final index, plotted in Figure 3.3, would then come to 1.08:
$$\frac{(2 \times 1.13) + (1 \times 1.05) + (1 \times 1.0)}{4}.$$

4

Barriers to Entry into the Overall Cost Leadership Strategy

The implementation of the overall cost leadership strategy by some Japanese firms has resulted in a change in the structure of the industry, and with that change a change in the height and nature of the barriers to entry. I mentioned above that the scale of output has increased substantially among the leading firms. Combined with the possibility of reaping benefits from economies of scale, price competition has taken on a much more important role in competition. In this chapter the nature and size of the economies of scale in the production and marketing of CNC lathes will be examined. Other barriers to entry have also become greater over the last few years. The speed of technical change has increased, and with it has arisen a greater need for R&D expenditure and skill formation. Furthermore, associated with the increase in the minimum efficient scale of production, there has been a growing need to increase exports. This has accentuated the need for access to a marketing network in foreign markets. Sections 4.1–4.4 therefore discuss the barriers to entry in the R&D process, in the procurement of components, in production, and in marketing. In section 4.5 the shape of the cost curve and the levels of output at which a firm reaches the minimum efficient scale of production are discussed. As mentioned in the previous chapter, there are alternative strategies to the overall cost leadership strategy. However, this is the dominant strategy and firms need to take its requirements as a point of departure for the formulation of their own strategy. This is the reason for the rather detailed elaboration on the barriers to entry associated with pursuing this strategy.

4.1   Research and development

The amount of money needed to be spent on R&D constitutes an important barrier to entry. This sum is partly a function of the *rate* of technical change, which can be measured using the change in lifetime of a basic design of a CNC lathe. This lifetime has decreased substantially since the 1960s. According to one of the leading Japanese CNC lathe builders, a design made in 1974–75 had a lifetime of eight years; while a design introduced in 1978 was being phased out in early 1983 and the expected lifetime of a model introduced in 1983 was approximately three years. Other builders in both Japan and Europe agree that the expected lifetime of a design put on the market in 1983 was around three years. This means that the pressure to expand expenditure on R&D is constantly increasing.

The larger firms in the industry, i.e. those following the overall cost leadership strategy, spend 5–7 million USD per annum on R&D. This can be contrasted with the annual *sales* of 15–20 million USD of the largest producers in Taiwan and Korea and with an annual production of less than 10 million USD of the largest Argentinian producer.

The cost of a new design varies, however, depending on the nature of the CNC lathe. A more *complex* one costs more to develop than a *simple* one. The degree of *newness* also affects costs. A copier is said to have lower costs than a firm designing a completely new model.[1] Finally, while firms generally compete on the basis of a series of CNC lathes, involving three to five models, a firm may well have several series in its product range. Thus, the size of this entry barrier depends on the strategy of the firm and cannot be analysed in general terms. It is true, however, that the large sums spent by firms pursuing the overall cost leadership strategy need to be taken into consideration by all firms in the industry.

How large, then, are the cost advantages for firms that can spread the initial R&D costs over a larger number of CNC lathes. The range of costs for developing a new series of CNC lathes is between 500,000 USD and 1,000,000 USD (interviews). Firms that base their development work on copying other firms' products tend to be at the lower end of the scale, whilst firms with a greater element of newness in their designs tend to be at the upper end of the scale. *Copying* firms also often produce simpler products, which reinforces their position at the lower end of the range. As was mentioned above, the lifetime of a design has declined to around three years. The total volume of output of that model or series is then a function of the rate of the output. If we assume that a firm sells only one series and that it produces 800 units per year, with a lifetime of three years the total volume becomes 2,400 units. If we then divide the initial costs of design development by this volume of output, the cost of R&D per unit of output would be 277–554 USD. However, if the annual sales are only 200 units, the R&D costs per unit of output would increase to 1,108–2,216 USD. (I have calculated with a 10 per cent interest rate on the initial R&D

expenditure. The figures thus represent an opportunity cost of using resources for R&D.) As the sales price of a CNC lathe may vary between 50,000 USD and 200,000 USD, the costs for R&D may appear very low as measured per unit of output.

If, however, we compare these figures with the firms' estimates of their annual R&D expenditures and with volume of output, the picture is a bit different. Thus, if we divide annual expenditure on R&D by the annual sales of CNC lathes, the largest firm would show a cost per unit of output of approximately 2,000 USD and other firms would be in a range up to 4,000 USD. These figures are closer to industrial reality for two reasons. First, some R&D projects result in models that are commercial failures. Secondly, some firms feel forced to design models that have a very limited market in order to supply a whole range of CNC lathes.

Thus, the advantages that larger firms may have are greater than if we consider only the costs associated with the development of individual successful series because they can spread the costs of all R&D over a large number of CNC lathes.

A barrier to entry of another kind is the skill requirements for designing CNC lathes. The introduction of electronic control systems has meant that the design process has become more complex. Disciplines other than mechanical engineering have entered the design task – such as electrical and electronic engineering, as well as microcomputer techniques and servo-techniques. As a result, lathes are no longer designed by an inventive mechanical engineer alone, but require a team with a multidisciplinary background. The market leaders' design teams have 30–50 per cent electronic engineers, out of a total of 150–275 designers.

Among the engineers, it is important to distinguish between core designers, who are the inventive *brains*, and the designers who perform design work of a more routine character. Without the core designers, the design work would come to a halt. There are also very few in any one firm. One leading European firm with three divisions claims that three to eight of the designers in each division constitute these core designers. All in all, this firm has 160–170 designers. A UK firm suggests that ten out of its forty designers constitute the core. The core designers embody the accumulated central knowledge of the design process and, even though they are few in number, they are the key to the success of the firm. These skilled designers are rarely found on the labour market and are therefore an extremely scarce resource for firms. For new entrants, these skills are all but impossible to buy in and therefore need to be developed inhouse through a long learning process. Of course, an increase in the rate of technical change also affects the demand for such designers. *Ceteris paribus*, firms have to increase the rate of skill formation in order to remain competitive.

4.2 Procurement of components

The dominant item in the structure of costs in producing (excluding marketing) a standardized CNC lathe is the bought-in components. Table 4.1 shows the cost structure for a CNC lathe and an engine lathe, both made by a producer based in a developing country in Asia.

Table 4.1. *Structure of costs for CNC lathes and engine lathes in one NIC-based firm (excluding sales costs and profits) (%)*

Cost item	Engine lathe	CNC lathe
Components	40	75
of which raw materials[a]	25	12
Blue-collar labour	10	4
Machining	25	8
Other[b]	25	13
	100	100

Source: firm interview.
[a] Casting, steel bars, etc.
[b] R&D, quality control, production control, etc.

In the case of engine lathes, the share of materials is only 40 per cent. The type of materials also vary greatly. In the case of engine lathes, raw materials account for 25 per cent of the production costs, out of which castings account for a sizeable share. Castings are generally produced in a labour-intensive way and often produced inhouse, although not in this particular case. In the case of CNC lathes, raw materials account for a very small part of production costs.

In the case of CNC lathes, bought-in materials account for 75 per cent of the production cost. Among the components, the dominant item is the control system, which includes the CNC unit and drives. These often account for up to 50 per cent of the cost of production, at least for smaller types of standardized CNC lathes. For more complex CNC lathes, the share may be around 20 per cent.

The CNC units are produced by large international firms in an oligopolistic industry. The scale economies are substantial as the main item of cost is investment in software development. Even though some producers of CNC units deny that price discrimination exists, it is conventional wisdom among machine tool producers in Japan that Fanuc, the main supplier of CNC units, has a different price for each customer. Price discrimination can take place because Fanuc has a degree of monopoly power, i.e. control over price (the source of which is probably lower average costs than its competitors), and the market can be segregated into various groups that are ready to pay different prices for the units. These firms have only imperfect information on the prices paid by other firms. Resale of the CNC units can be controlled as Fanuc services and maintains these units. In the machine tool business (especially in small firms), one can hear that 'the price of a CNC unit covers the cost of producing three units'. Even though the cost probably refers to marginal cost, it still gives an indication of the magnitude of price discrimination. The rationale, for the supplier favouring larger customers is, for example, that they enable the firm to produce on a very large scale and a stable relationship with a large machine tool firm creates a quasi-captive market for the supplier. The cost of switching the type of CNC unit is high for the machine tool builder as his customers need to learn about another type of CNC unit. Stable relationships also tend to reduce the risks in product development for the CNC supplier. Price differentiation would appear to exist in Europe too.

The basis for the setting of prices is the bargaining process. The suppliers have official price lists, but even one of the largest suppliers of CNC units in Europe conceded that it has no fixed prices but that the price was subject to wide fluctuations depending on the bargaining strength of the buyer. However, a one-to-one relationship between the number of units demanded by the buyer and the degree of price reduction does not exist. According to interviews with a number of Asian CNC lathe builders, *special relationships* are sometimes established between a certain supplier of CNC units and some user firms. A special relationship can be established, for example, with the producer that initiates the production of NCMTs in a country and thus sets a standard for imitators. It would also appear as if governments can intervene in the bargaining process by allowing local production of the CNC unit and thus create a special relationship. One firm with such a special relationship and which currently produces fewer than 200 CNC lathes per annum, pays approximately 40 per cent less for its CNC units and motors than another firm that produces approximately 50 units per year. As is shown in Tables 4.2 and 4.3, we would expect a firm to have to acquire something in the order of 1,000 units per year in order to get such a large discount. The benefiting firm claims that it pays about the same as one of the largest Japanese producers, which produces about 2,500 units per year, owing to its *special relationship* with the supplier. It would thus appear that this supplier of CNC units judges the growth potential of the buying firms and tries to secure future markets for its CNC units by favouring the most *progressive* firms.

Table 4.2 shows the estimated prices of CNC units for firms buying various yearly volumes, as suggested by a larger supplier of CNC units. A leading European producer of CNC lathes corroborates these figures; the estimated reductions in the price of the CNC unit and the drives bought by this firm are shown in Table 4.3.

Table 4.2 *Price advantages as the annual level of demand for control systems (incl. motor) increases as suggested by a large supplier*

Annual demand	Normalized prices
50	1.0
300	0.8
700	0.7
2,000	0.5

Source: interview with a representative of a supplier of control systems.

Table 4.3 *Price advantages as the annual level of demand for CNC units increases, as suggested by a leading European producer of CNC lathes*

Annual demand	Normalized price
1	1
100	0.8
500	0.7
1,000	0.6

Source: interview.

Of course, for many machine tool producers, the relationship with the supplier of control units is a very unequal one. For example, Fanuc produced 15,681 CNC units in 1981 and the largest NIC-based producer buys around 200 CNC units per year. However, there appears to be a trend towards growing competition in the control field. This refers partly to the availability of cheap control systems from other specialized producers, e.g. General Electric, Siemens, Olivetti, Electronic Control Systems, etc.

As was mentioned in section 3.6, a number of machine tool builders have also integrated backwards. For example, a leading Japanese machine tool firm now has an inhouse production of 3,000 units per annum, whereas it previously bought control units from Fanuc. In Chapter 3 we also saw that leading German producers have integrated backwards, including the largest German producer, Gildemeister. Hence, the competitive situation appears to be changing, which implies not only that there is an upper limit to the degree of exploitation by the control unit supplier but also that firms like Fanuc may have a greater interest in developing alternative markets, e.g. dynamic machine tool builders in the NICs.

The price reductions attainable for the other bought-in materials are less than for the control units. It was estimated[2] that for the mechanical and hydraulic components, such as the ballscrew, the chuck and the hydraulic system, the maximum price reduction is 20 per cent. Finally a 10 per cent price reduction can be obtained in the procurement of raw materials, such as castings and steel bars. In Table 4.4, the maximum price reductions are weighted with the share of these items in the total cost of production (excluding marketing) for firms producing at low levels of output, say less than 100 units per annum. Obtaining these maximum price reductions would, for a large-scale producer, mean a reduction of approximately 23 per cent in total costs as compared with a small-scale producer. A key factor determining the price competitiveness of CNC lathe production is, therefore, a good relationship with component suppliers.

Table 4.4 *Estimated maximum price reductions attainable for components (%)*

	Cost item			
	Control unit %	Mechanical components %	Raw materials %	Total %
The share of components in the cost of production[a] at output levels below 100 units p.a.	47	17	6	70
Maximum price reductions attainable for components	40	20	10	
Resulting reduction in costs	18.8	3.4	0.6	22.8

Source: Elaboration of interviews with firms.
[a] Excluding marketing costs.

4.3 Manufacturing

The process of manufacturing a CNC lathe includes the basic processes of metalcutting and assembly. To these can also be added inspection, quality control, production planning, general administration and design work. The design function has been amply discussed already, and will not be further treated here.

The machining process of the machine tool industry has undergone major changes in the past decade. Machining centres, which are a combined milling, drilling and boring machine, are used, practically speaking, everywhere. The firms in the NICs also use this technology extensively. Furthermore, in the past few years there has been a diffusion of smaller and larger systems where the material handling is automated. The simplest form of system is a machine centre equipped with an automatic pallet-changer (FMM) and some extra pallets. Larger systems involve, for example, several machining centres, all with automatic pallet-changers and where the work piece is transported by an automatic guided vehicle. Such systems are often called Flexible Manufacturing Systems (FMS) (see Edquist and Jacobsson, 1985a). Very significant productivity increases can be gained with the adoption of FMMs. Although the use of FMMs is certainly not restricted to larger firms, the larger systems are predominantly installed by larger firms, e.g. those Japanese firms following the overall cost leadership strategy. Interestingly enough, however, a smaller UK company, which was recently bought by a large US machine tool firm, has installed a very large FMS system. A smaller Korean firm is also installing an FMS system (Edquist and Jacobsson, 1985b). It is not easy to specify the benefits to the user of installing a system of some type as they depend on so many factors, for example the technology with which one compares it. Compared to *conventional manufacturing*, Boston Consulting Group suggests that a 40–50 per cent reduction in total production cost could be achieved (Boston Consulting Group, 1985, p. 32). Since total production costs also include assembly and testing, this figure is most likely an overstatement. I would rather suggest, at most, a 35 per cent cost reduction when the comparison refers to the use of stand-alone machining centres and the figure relates only to the machining process.

As for labour in the assembly process, Adam Smith pointed out the organizational determinants of the efficiency of labour. In contrast to the high divisibility of labour, organizational forms are less divisible. Larger firms can implement a work organization based upon more specialized tasks and can therefore draw more benefit from the economies of labour specialization. A possibly extreme example in this industry is the introduction of specially constructed assembly lines for the best-selling models of CNC lathes by two Japanese builders. While one Japanese firm claimed that labour productivity could be doubled, other firms were more modest and suggested that the maximum increase in productivity was 40 per cent.[5]

Finally, the overhead costs (e.g. design, administration, production planning) are largely fixed costs and it is therefore considered that a 70 per cent reduction in overhead costs per unit of output is feasible.

4.4 Marketing and after-sales services

The costs for transport, sales and after-sales services account for a considerable part of the sales price. Smaller firms selling standardized products commonly claim that about 30 per cent of the customer price covers these costs. The composition of these costs varies between firms, but as a rough approximation it seems correct to say that 20 per cent goes to the final dealer, 5 per cent to repair and maintenance and 5 per cent to cover transport.

Owing to the very high costs involved in developing knowledge about foreign markets, in foreign and geographically remote domestic markets firms normally sell via local independent distributors. Direct sales to the customer are more frequent in the home market.[4] Firms that initiate exports to a given market therefore normally employ a local distributor, who either sells directly to the user or to another distributor. It is common for the machine tool producer to have the distributor also perform the repair and maintenance function. For a newcomer, *access* to a local distributor network may be an absolute barrier to entry. It is in the interest of the distributor to sell products that are popular, thus reducing the sales costs per unit. As newcomers are less well known, they may face the problem that the distributor does not promote their products enough. One Japanese firm tries to remedy this situation by offering a higher mark-up for firms that put more emphasis on its products in their sales efforts. Other firms may simply say: 'Give us what you can get in the market'. In other words, a distributor has to be given a good incentive to take the risks associated with the investment that is needed to create a name for a new entrant. This effort has to be supplemented by massive advertising and participation in trade fairs, etc. Finally, the price to the customer has to be very low initially; one Taiwanese firm suggests that their lathes need to be priced 30 per cent lower than an equivalent Japanese lathe in order to sell. Thus, breaking into a market involves overcoming an absolute barrier to entry in the form of access to a marketing network and high costs in terms of creating a name or a reputation.

When a firm has begun to penetrate a market it is usual to set up an office for direct sales to customers and/or for selling to regional distributors. This service centre often fulfils the function of repair and maintenance, as well as stocking final products and spare parts. In order to be able to sell CNC lathes, the buyer must be assured of good service and maintenance; delivery time is also important in this respect. The availability of a service centre is often seen as another absolute barrier to entry for firms in some larger markets.

The fixed costs for setting up such a centre vary, depending on the firm's ambitions. The largest producer of NCMTs in the developing countries recently invested 3–4 million USD in a US service and sales centre employing sixteen to seventeen service engineers and seven salesmen; 50–60 per cent of the costs involved were used for building up a stock of products and spare parts. Another developing-country firm claimed that the cost of setting up a sales and service centre in Europe was prohibitive while demand was low.

There are important scale economies to be reaped as the stock of a firm's machine tools increases in a particular market. Scale economies come predominantly from a reduced need for selling effort. This stems from the fact that *repeat buying* constitutes a large share of the market; that is, firms that already have a CNC lathe from a particular producer and are pleased with it take the initiative themselves to buy new ones. The first sale of a machine tool to a customer, in contrast, generally speaking involves a much greater sales effort. Hence, as the stock of a firm's machine tools increases in a market, the greater is the share of repeat sales in its total sales.[5]

Another source of scale economies lies in the fact that with a higher density of that particular firm's machine tools in the market, more customers can be served by one service engineer.[6] (Normally, the producer of the CNC units takes care of the service and maintenance of that component, while the lathe producer takes the responsibility for the rest of the machine tool.)

The reduction in marketing costs per unit of output as the stock of units increases in a market naturally encourages the producer to try to reap the benefits. The producer can do so by taking over a larger share of the marketing effort. In some cases this even entails building up a special organization. For example, the firm that produces the largest number of CNC lathes per year in the world, Mori Seiki, has eleven regional offices in the US market. The trend of Japanese machine tool builders to shift from relying on trading companies to developing their own distribution channels may be the result of an attempt to benefit from such economies of scale.

The exact size of the scale economies in marketing and after-sales service is extremely difficult to ascertain. One medium-sized builder suggested that the share of sales and after-sales service in the customer price may be as high as 50 per cent when only a few units are sold per annum, whilst the percentage may be reduced to about 20 when 50 units are sold. For larger firms, this cost is even lower; the minimum cost is probably about 15 per cent of the sales price.

About 5 per cent of the price paid by the customer pays for transport. Of course, the cost of transport is generally higher the further away the producer is from the market. As in the procurement of components, firms can bargain about the cost of transport, even though the scope for price reductions is lower – one large Japanese firm claims that a reduction of 5–7 per cent can be made by large firms (interview). Also, at very low volumes, the official prices can vary a lot with scale. One distributor in Europe claimed that the price per unit of shipping from Asia could be reduced from 8 per cent to 4 per cent of the f.o.b. price if a container including two–three CNC lathes were bought instead of one CNC lathe (interview).

This discussion of the nature and size of economies of scale in the production and sale of standardized CNC lathes is summarized in Table 4.5. All in all, the maximum scale economies would seem to be around 41 per cent.

Table 4.5 *An estimate of the size and origin of economies of scale in CNC lathe production (%)*

Item	Share in original costs at production levels of around 100 units p.a.	Maximum cost reduction	Effect of maximum cost reduction on original total cost
Components	49		
Control unit	33	40	13.2
Mechanical components	12	20	2.4
Raw materials	4	10	0.4
Machining	9	35	3.1
Assembly	4	40	1.6
Overheads	8	70	5.6
Marketing, sales and after-sales costs	30	50	15.0
Total cost	100		41.3

Source: Estimations based on firm interviews.

4.5 An attempt to specify the minimum efficient scale of production

After ascertaining the magnitude of the scale economies involved in the production and sale of CNC lathes of a standardized type, it is of course of interest to discuss the shape of the cost curve as the scale of output increases. For a new entrant it is vital to know if the bulk of the benefits of economies of scale are reaped at a relatively low level of output or at a large level of output. In other words, where does the break-even point lie? (I underline again that the discussion refers to firms following the overall cost leadership strategy.)

The break-even point is also, however, a function of a number of factors other than scale economies. Such factors as labour costs, capital costs, output mixes and x-efficiencies have to be taken into account by the individual firms. Furthermore, as was discussed in section 4.2, there is no one-to-one relationship between the size of a firm and the price that it pays for its components. We also saw in Chapter 3 that an Italian firm following this strategy gained a competitive edge through access to cheap CNC units from being owned by Olivetti. Certainly, this reduces the break-even point for this firm in a substantial way. Hence, there is by no means just one cost curve that is applicable to all firms.

A further factor of importance for a new entrant is the behaviour of the leading firms with respect to profits. When the leading firms set a price that is higher than average costs, and thus make a profit, entry is easier for a new firm. As was shown in Table 3.10, the leading Japanese firms have made large profits in the 1980s.

In spite of these problems, a few words can be said about the break-even point. One way of determining the magnitude of the volume of production that is necessary in order to become competitive is to look at the characteristics of the firms that have survived the recent structural changes in the industry. Of the seven firms pursuing the overall cost leadership strategy, three produce more than 1,500 units per year while three produce 800–1,000 units. The average profits of the three leading firms in 1981 approached 20 per cent, while the average of the three other firms was around 3 per cent. The order of magnitude has not changed since then. Hence, it would appear that, in the developed countries, the break-even point lies between 800 and 1,000 units per year. As mentioned in Chapter 3, however, the Italian firm following this strategy produces only 350–370 units per year, although, as pointed out above, it gains a lot from being part of the Olivetti group. Furthermore, import penetration in Italy is very low for Japanese machine tools, which suggests that the pressure on prices could be lower than elsewhere.

Another way of approaching the problem is to examine the sources of scale economies and see at which point most of these are exhausted. As was shown in Table 4.5, three items account for most of the economies of scale. These are the price of control units and overheads and marketing. It can be seen from Tables 4.2 and 4.3 that the bulk of the price reductions for control units are attained at volumes of less than 1,000 units, and that a large price reduction can be obtained for volumes of 500–700 units. Overheads can largely be considered as a fixed cost. This means that the cost per unit of output declines fastest at quite low levels of output. Finally, the main economies of scale in marketing can be reaped when a sales and service organization has been established in a market. It should be noted that these markets are often confined to a limited geographical area owing to the concentration of industry. The markets can therefore to some extent be treated as independent from one another. It would thus appear that fewer than

100 units per year need to be sold in order to reap the bulk of the benefits of economies of scale in a given, regional, market. Indeed, Mori Seiki, the largest producer of lathes in the world, sold around 150 CNC lathes in the UK in 1979 (*The Engineer*, 1979).

Thus, both ways of approaching the problem point to an exhaustion of the main sources of scale economies at levels of output below 1,000 units per year. In fact, the main scale economies may be exhausted at production volumes of 500–700 units per year. This estimate would seem to be in line with that of the Boston Consulting Group:

> . . . an annual output of 400 machines is the absolute minimum scale necessary for survival. If one assumes that a European supplier can command a modest price premium in his local market, and adds in the higher transport and the tariff costs incurred by his Japanese competitors, he could probably break even at this scale. [Boston Consulting Group, 1985, p. 33]

However, for firms to be viable in the long run they need to perform better than being close to or at the break-even point. I would therefore estimate that in the long run firms need to approach a yearly volume of towards 1,000 units. Again, it should be emphasized that for the individual firm a number of other factors, mentioned above, need to be taken into consideration in ascertaining a specific cost curve.

I have analysed the nature of competition in the global CNC lathe industry and the character of the barriers to entry into it, and in particular the strategic group of overall cost leadership. In the next chapter I proceed to analyse the position of the NICs within the CNC lathe industry.

Notes

1. This refers to the mechanical parts. Copying is extremely difficult to undertake in the case of the CNC unit as the technology cannot be observed in the same way.
2. These estimates were based on interviews with several firms.
3. In this context it is of interest to note that several Japanese firms interviewed saw labour as more of a fixed cost than a variable cost. This would seem to be a function not of a lifetime employment system, as these firms have made workers redundant in times of crisis, but rather of the ability to increase labour productivity drastically as and when it is required. Firms 1 and 5 in Table 3.10 claim that they have kept the number employed constant since 1977–78. At the same time, the value of production increased by more than 300 per cent. To some extent, this increase in labour productivity can be attributed to a change in product mix, but the management also claims that the motivation of the labour force is extremely important for attaining productivity increases (interview).
4. For example, one source suggests that, in France, whilst 20 per cent of imported CNC lathes are sold directly to customers, the percentage increases to 50 for sales by domestic producers (Planning Research & Systems Limited, 1979, p. 40).

5 The general fact of repeat sales can be illustrated by the experience of one of the pioneers in CNC lathes in Japan. The firm claims that in 1967–68 it sold 51 units to thirty-six customers. In 1982, twelve out of these thirty-six customers had a stock of over 20 units of this firm's CNC lathes (interview).
6 Some Japanese firms claim that at least 100 machines need to be installed in one market before it is worthwhile setting up a local service and maintenance staff (interview).

5
The Position of the NICs within the CNC Lathe Industry

The machine tool industry is a fairly sizeable industry in the NICs. Table 5.1 indicates the value of production, trade and demand for machine tools in a number of NICs. The rank of each country in the world is also indicated. It may be noted that, although these countries have not yet reached a position where production exceeds demand, the discrepancy is not that great. It may also be noted that several of these countries are large in terms of markets and production of machine tools. For example, machine tool demand in India was the eleventh largest in the world in 1984, while Taiwanese production ranked as number fifteen in 1983.

Table 5.1 *Production, exports, imports and demand for machine tools in five NICs, 1983*

Country	Production (US$ m.)	World rank	Exports (US$ m.)	Imports (US$ m.)	Demand (US$ m.)	World rank (1984)
Argentina	28	29	14	24	38	n.a.
Brazil	98	24	24	44	118	n.a.
India	184	17	22	149	311	11
Korea	119	22	21	140	238	20
Taiwan	211	15	135	113	189	19
Total	640		216	470	894	

Source: NMTBA (1985/86). For Korean trade figures I have used Office of Customs Administration (1983).

In all of these countries, some leading lathe producers are attempting to move into the production of CNC lathes. Data of the production and demand for CNC lathes are given for the five countries in Table 5.2. It is evident that the Asian NICs, Korea and Taiwan, are the most dynamic countries, with respect to both the production and the diffusion of CNC lathes. Even these two relatively successful countries are small, though, in terms of their share of world production of CNC lathes. In 1984, Korea and Taiwan had 6.9 per cent and 7.2 per cent respectively of the world production[1] of conventional lathes (in value). Their shares of the world production of CNC lathes were only 0.7 and 0.9 per cent respectively.

Table 5.2 *Production of and demand for CNC lathes in five NICs (units)*

Country	Production		Demand	
Argentina	10	(1983)[2]	45	(1981)
Brazil	120[b]	(1982)	150[b]	(1982)[c]
India	15[b]	(1983)	101[b]	(1983)
Korea	268	(1984)	248	(1984)
Taiwan	347	(1984)	250	(1984)

Sources: Argentina: Center on Transnational Economy (1984); Brazil: Rattner (1984); India: production data – Indian Machine Tool Manufacturers Association; demand data – Edquist and Jacobsson (1986); Korea: Table 8.10; Taiwan: data supplied by MIRL.
[a] 11 months.
[b] All types of numerically controlled machine tools.
[c] Assumed no exports.

Data at the firm level are available from Argentina, Korea and Taiwan. The firms in these countries comprise a fourth strategic group within the international CNC lathe industry. Their overall strategic position can be described as one of pursuing a *low-performance strategy*. (In the strategic map in Figure 3.3, four firms from these countries were also plotted. They were found in the southwest corner of the figure.) I shall describe the low-performance strategy in terms of the seven decision variables or characteristics of strategic groups listed in Table 3.22. The strategy pursued by the NIC-based firms involves focusing on users (target groups) that do not require a high cutting capability or high precision and that are extremely price sensitive. Typical users can be small subcontractors, first-time users of CNC lathes, schools and metalworking plants in the NICs. The fact that the CNC lathes are of standard design and of lower performance, given their size, than other CNC lathes, means that they are cheaper to build and the price can therefore be kept low. This is achieved, for example, by using motors with a lower horsepower and castings of less rigidity. The strategy is also less demanding in terms of design skills (R&D)

both because of the nature of the product and because these firms can, and very often do, copy other firms' models. Copying also implies lower R&D costs. Given the standardized nature of the product, independent distributors can be used. Access to a distribution network is facilitated as they can use a network that sells CNC lathes of firms pursuing a different strategy. For example, in Sweden, the leading Taiwanese producer of CNC lathes uses a distributor that also sells CNC lathes of a German firm producing high-performance CNC lathes. Finally, all the firms produce a small number of CNC lathes per year. The characteristics of the low-performance strategy are summarized in Table 5.3 using the same terminology as in Table 3.22, where I summarized the characteristics of the other three strategic groups in the world industry.

Table 5.3 *Summary of the main characteristics of the low-performance strategy*

Characteristics	Low-performance strategy
Product:	
Degree of standardization	Standard
Performance	Very low–low
Price	Very low
Target groups	Highly price-elastic small firms
Marketing	Mainly through independent dealers
R&D	Low cost, frequent copying of designs
Volume	Low

There are a number of problems for firms following this strategy. First, in terms of the requirements set by the international market, nearly all of the firms produce below the minimum efficient scale of production. As was seen in Table 4.5, the main sources of scale economies lie in the purchasing of components and in pre- and after-sales service. These sources also apply to the low-performance strategy. Indeed, none of the eight firms interviewed claimed that they made a profit on their production of CNC lathes. A contributing factor to these losses is the costs associated with creating a brand name for new entrants, which involves not only large expenditure on promotion but also the pursuit of a low price profile for a long time period. This position is however untenable in the long run on account of the existence of economies of scale. Secondly, it is very questionable that the world market for such low-performance CNC lathes is sufficiently

THE POSITION OF THE NICs 105

large to absorb the output of these firms if they were to produce volumes large enough to reach a break-even point. A move to another strategy that involves not only larger volumes of production but also a technically upgraded product would therefore seem to be required. In spite of these problems, the low-performance strategy is a way for firms to enter the industry and gain experience. It should, however, be seen as only the first step in a longer process of consolidation of their entry. After describing the position of eight NIC-based firms within the low-performance strategy, I shall further discuss the direction in which these firms will have to move in the medium and long term in order to have a viable production of CNC lathes.

5.1 The position of eight NIC-based firms within the low-performance strategy

The position of the firms *within* the low-performance strategy varies, however, as do the possibilities for the various firms to consolidate their entry. I shall therefore analyse the individual firms' positions within the low-performance strategy.

In Chapter 4, I discussed several barriers to entry – economies of scale, technological capabilities or design skills and access to marketing networks and associated brand image. In Table 5.4 selected characteristics of eight NIC-based firms producing CNC lathes are presented; one firm is Argentinian, two are Korean and five are Taiwanese.

The first characteristic is the *sales volume*[1] for numerically controlled machine tools. This indicates to what extent scale economies can be reaped. The next five characteristics are all indicators of the design skills in the firms.

The *type of design*[2] can vary tremendously. When numerical control began to be applied to lathes (in the mid-1960s), the normal method was to take a conventional lathe and more or less just add the control unit to it. Thus, the design concept did not differ from that for conventional lathes. As one reports says:

> In the early days of numerical control, it was not uncommon to fit NC systems to existing machine tools as distinct from machine tools designed and made for use with NC. *Retrofitting*, as this is called, did produce some successful NC machines but it also produced many unsatisfactory ones. The mechanical characteristics of conventional machine tools made prior to 1955–1960 – stiffness, frictional characteristics of slideways, and feed drive systems – were often unsuitable for use with NC and retrofitting of this kind is now unusual

Table 5.4 Selected characteristics of eight NIC-based firms producing CNC lathes

Firm	Country	(1) Sales of CNC machine tools (units)	(2) Type of CNC lathe	(3) Origin of design	(4) No. of models of CNC lathes[a]	(5) No. of design engineers	(6) Ratio of design engineers to total employees %	(7) No. of electronic engineers[b]	(8) Marketing network in the US
A	Taiwan	240 (1984)	modern	own	3[c]	35	5	5	yes
B	Korea	154[h] (1983)	modern	own	4	60	9	12	yes[e]
C	Taiwan	138 (1984)	modern	copy	2[d]	20–25	3	5	yes
D	Korea	68 (1983)	modern	own?	2	30	4	10	yes
E	Argentina	150–200	10[f] (1983)	old + modern	own + licence	several[g]	7	4	1	no
F	Taiwan	250	6 (1982)	modern	copy	2	5	2	2	yes
G	Taiwan	90	16 (1982)	very old	own	1	5	5	2	yes
H	Taiwan	250	5 (1982)	old	licence	1	6	3	1	yes

Sources: Firm interviews and KMTMA (1984), p. 5 and (1985), p. 11, apart from columns (1) and (5) in Firm E, for which the source is Center on Transnational Economy (1984).

[a] Firms A, B and C also have older CNC lathe designs of own design or copies. The numbers here refer only to their modern designs.
[b] These are included in number of design engineers (column 5).
[c] The firm also has four machining centre designs.
[d] The firm also has two machining centre designs.
[e] The marketing network belongs to the large engineering conglomerate that owns the machine tool firm.
[f] In the first 11 months.
[g] The firm has two old designs of own make and licenses an unknown number of models from a Japanese firm.
[h] In the first 7 months of 1985 this firm produced 145 CNC machine tools.

except for some of the simpler and cheaper NC systems. [*Metalworking Production*, 1979, p. 157; emphasis added]

With the advent of the minicomputer, the CNC lathes began to a greater extent to be redesigned to utilize fully the advantages of electronics. The first CNC units were supplied around 1970. However, as late as 1975 a report in *The Engineer* claims that: 'Nearly every lathe maker now offers numerical control, though fewer have machines designed from the outset for NC' (*The Engineer*, 1975, p. 44).

The type of CNC lathe common around 1975–77 was a large and very versatile machine that could handle a large variety of different sizes of components. The bed was normally flat.

By the end of the 1970s specialization had taken place whereby smaller CNC lathes were developed. The bed is also of a slant type indicating that the chips are more easily removed. This development was a response to the fact that not all customers need a very versatile lathe in terms of the size of components that can be turned. It was also a response to the cheaper CNC units, which made it economical to apply CNC to smaller and therefore cheaper lathes too. These slant-bed CNC lathes are normally more difficult to design and produce than the large flat-bed type. In terms of markets, the various types of designs have to a large extent substituted for each other, even though the time period involved has been long and some overlapping has taken place. There is, for example, still a market for the large flat-bed CNC lathes even though it is marginal compared to that for the slant-bed CNC lathes.

Thus, in terms of the type of designs produced by the various CNC lathe manufacturers based in developing countries, we can distinguish between a retrofit, which is the easiest to design, a large flat bed, which is more difficult to design, and a smaller slant bed, which is the most difficult to design. I call these design concepts 'very old', 'old' and 'modern'.

It is possible to differentiate between the *origins*[3] of the design: licensed, which incidently applies to only one firm; copied, which is very frequent; and own design. It is not unusual that a firm that has produced its own design for the 'very old' or 'old' type feels that it has to rely on copying when it wants a modern design. Apart from the Argentinian firm, none of these firms have licensed a modern design, partly because they claim that these are not available. (The Argentinian firm recently entered into a licence agreement with a Japanese firm; however, the design was five years old.)

The *number of models*[4] refers to the fact that firms that have established themselves in the industry compete with a whole series of CNC lathes, all based on the same basic concept but varying in size. A larger number of models indicates a greater mass of design capabilities.

The *number of designers*[5] is an important indicator, not only of the mass of capability in the firm but also, and perhaps most importantly, of the emphasis that the firm places on generating skills. All these firms are definitely in the learning phase and they all need to develop the core designers discussed in section 4.1. As electronics has now entered the design phase, the number of *electronic engineers*[7] in the team is also indicated.

The last characteristic is whether or not a firm has begun to develop its own *marketing network*[8] by establishing a service and sales centre in the USA, the largest market in the world. Because domestic demand (as shown in Table 5.2) is limited in relation to the magnitude of the scale economies involved, there is a great need for exports, at least in the absence of protective trade barriers.

Based upon this information, a map has been drawn to indicate the position of the firms within the low-performance strategy (see Figure 5.1). The two most important characteristics of the nature of competition are scale economies and design capabilities. The number of NCMTs produced per annum has therefore been put on the horizontal axis and a measure of technological capabilities on the vertical axis. The further to the northeast that a firm is placed on the map, the better are its chances of consolidating its entry. Of course, other factors also affect the outcome, such as the institutional affiliation of the firm and government policies, but these can be treated as external factors at this point. The map gives a picture of the firms' situation today and does not show the dynamics of the firms, which are more affected by the two factors mentioned above.

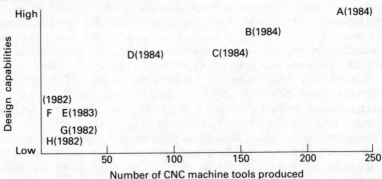

Figure 5.1 *Map of the position of eight NIC-based firms within the low-performance strategy*
Source: Elaboration of Table 5.4.

Evaluating the importance of the five factors that indicate the technological capabilities of the firms is problematical. Two groups of firms are however immediately recognizable. Firms A–D all produce modern CNC lathes; have more than one model; and have a substantially larger number of designers, including electronic engineers, than the other firms. Within this advanced group, it would seem appropriate that firm A should be ranked as the most advanced as it has its own, modern design not only of three CNC lathes but also of four machining centres. Firm B is placed second although it has developed a series of CNC lathes of its own design, since its own designs relate only to CNC lathes and not to its machining centres. Firms C and D are placed below firm B as they have substantially fewer design engineers. These two firms are broadly equal in design capabilities.

The gap between these four advanced firms and the others would seem to be very substantial, not only because most of the others produce the old types of CNC lathes (in very small numbers) but also because they have only very few designers. Among these firms, firms E and F would seem to have an advantage over the others. Both firms have two models of their own design and firm F also produces a modern design, even though the basic concept is a copy.

From this map, it may be concluded that firms A–D have the greatest chance of consolidating their entry because they are both the strongest in terms of design skills and have the largest volume of production of CNC machine tools. Indeed, firms A, B and C can be said to have consolidated their entry into the low-performance strategy, in that it makes sense for them to begin thinking of a shift to another strategic group with a better long-term viability. Firm D is in the process of consolidating its entry. The remaining firms exist on the periphery of the industry. Hence, the shift to another strategy, which is judged to be required in the medium or long term for all firms in this strategic group, can be of immediate interest only to the three leading firms and for firm D in the more distant future.

The question then arises of which of the three strategies can be judged to be the most appropriate for these firms to follow in the medium or long term. In section 3.2, I argued that the lower the entry barriers are to a specific strategic group, or the greater the similarities between the would-be entrants' resources and those of the firms in the strategic group, the lower are the risks involved in changing strategy and entering the group. We would then need to compare the resources of the leading NIC-based firms with those of the firms in the three strategic groups in the international industry.

Although the minimum efficient scale of production in the strategic group of the *overall cost leadership strategy* is estimated to approach 1,000 units per annum, the lower labour costs in the NIC can reduce this figure somewhat. Indeed, the leading Korean firm estimates that it will break even at an annual volume of production of 350 CNC lathes.

This figure is, however, somewhat below the *normal* break-even point as this firm has access to very cheap control systems.[2] Shifting to the overall cost leadership strategy would also imply an upgrading somewhat in the technical performance of the CNC lathes as well as producing a larger number of models. This process has already been begun by the two leading Asian NIC-based firms. As was noted in section 4.1, the firms pursuing the overall cost leadership strategy have 150–275 designers. The two leading NIC firms have 35 and 60 designers respectively. These firms will therefore probably need to at least double the number of designers in the medium term in order to be able to keep up with the rapid product development as well as to be able to supply a series of models. On the other hand, the target groups would be very similar to those of the low-performance strategy, as would the marketing organization used, namely independent distributors.

As for the *focus* and *differentiation* strategies, one needs to first point out that the local demand for the type of CNC lathes produced by firms following these strategies is very limited in the NICs. Thus, if the CNC lathe producers were to follow either of these strategies, they would need to export the vast majority of their production. Penetrating the Western markets for these types of CNC lathes would necessarily involve acquiring a reputation for high quality and reliability, as well as finding a marketing network with direct links into some larger firms. Such a marketing network is a prerequisite for following the differentiation strategy, and the leading edge firms also constitute a significant share of the market for the firms following the focus strategy. Brand names and marketing networks constitute in these cases formidable barriers to entry. Furthermore, the NIC-based firms would need to change their product drastically if they were to follow either of these strategies. The product would need to be changed from low performance to high performance, and would often need to include elements of custom design as well as system developments. This would require a very considerable discontinuity in design skills.

Thus, the only similarity between the low-performance strategy and the focus and differentiation strategies is the volume of output. In all other characteristics, the divergence is very large. On the other hand, the overall cost leadership strategy and the low-performance strategy show important similarities in the products, the target groups and the marketing organization used. The discrepancy here lies mainly in the scale of output. Furthermore, whilst the NIC-based firms would need to double the number of designers if they were to follow this strategy, they would nevertheless not face a need for a large discontinuity in the level of skills of the individual designers and the orientation of their design work since the character of the CNC lathes produced by firms in the two strategic groups is rather similar.

All in all, I would judge that the greatest similarities exist between the low-performance strategy and the overall cost leadership strategy and that it is this strategy to which the NIC-based firms will ultimately need to change if they want to consolidate their entry into the production of CNC lathes. None the less, the risks of shifting to the overall cost leadership strategy are very considerable since the leaders all have volume advantages and are solid in financial terms (see Tables 3.9 and 3.10). Furthermore, the main market expansion of CNC lathes has probably already taken place.

I have analysed the nature of competition in the global CNC lathe industry, the position of the NICs in this industry and the strategic reorientation that the leading firms will eventually need to undertake.

In the next three chapters I shall analyse the experience of lathe producers in the three NICs chosen for field work: Argentina, Taiwan and Korea. The objectives of these chapters are (1) to understand how different contexts, in particular governmental policies, in the three countries have influenced the growth of resources in the leading lathe producers, and (2) to understand how the present government policies influence the growth of resources within these firms in the field of CNC lathes.

The presentation will follow basically the same structure in each chapter. In the introductory section, I shall broadly paint the overall development strategy of the country and the growth performance of not only the whole economy but also the manufacturing sector and especially the engineering sector. Next, I shall discuss the development of the machine tool industry and the policy context of its development. I shall then proceed to analyse the size of the domestic market for CNC lathes. The analysis is more extensive in the case of Argentina than in the cases of Korea and Taiwan, the main reason being that I began my study in Argentina and I can to some extent transfer the results from Argentina to the other countries.

Next, I analyse the historical development of the leading CNC lathe producers, their present position in the CNC lathe industry and the main strategic issues facing them today. I shall also discuss the role of current government policy in the choice of strategy by the firms. In the final section, I shall summarize the main conclusions.

Notes

1 World production is defined as the output of Korea, Taiwan, the USA, Japan, the FRG, France, Italy and the UK. The six OECD countries accounted for 85 per cent of the non-socialist world's output of machine tools in 1981 (*American Machinist*, February 1983).

2 Upon being asked why they pay so little for the control systems, the director of the firm exclaimed, 'That is politics' (interview). In this connection, it can be mentioned that the supplier of the CNC units, Fanuc, has been permitted to establish a joint venture in the Changwon industrial complex where this machine tool firm is located.

6

The Case of Argentina

This chapter presents the case study of Argentina. In section 6.1, I shall paint a broad picture of the historical development of the industrial sector, and in particular the engineering sector. The engineering sector is fairly sizeable as a consequence of the inward-oriented policy that has been pursued during the past twenty years. The development of the Argentinian machine tool industry within this overall development pattern is described in section 6.2, while the policy of the government vis-à-vis the engineering sector and the machine tool industry is discussed in section 6.3. The governmental trade and exchange rate policies that have induced inward-looking behaviour in the machine tool industry are outlined. In section 6.4, I analyse the present and potential market for CNC lathes in Argentina and conclude that it is of very marginal size in relation to the minimum efficient scale of production.

In section 6.5 I turn to the only producer of CNC lathes in Argentina and analyse in detail both its present resources and the resources required to compete internationally. I conclude that the strategy pursued by the firm in the past has not enabled it to accumulate sufficient resources to be able to compete internationally in the near future. Rationally, the firm has therefore chosen a regional market strategy behind tariff barriers. The main results are summarized in section 6.6.

6.1 Growth and structure of the engineering industry

The Argentinian economy is fairly sizeable: in 1983, the GDP was over 71 billion USD, which was the same as the Korean economy in that year (World Bank, 1985).

Argentinian manufacturing production grew at a rate of 5 per cent per annum between 1950 and 1979 and by 1960 manufacturing accounted for 28 per cent of GDP, a figure that rose to over 33 per cent in 1975 (United Nations, 1980a). By 1983, it had however declined to 28 per cent (World Bank, 1985).

The growth of the industrial sector in Argentina has been of a very cyclical nature. As one source notes, between the years 1955 and 1977 the growth rate exceeded 10 per cent for six years and was zero or negative for six years. The same cyclical fluctuations were also found in investment in machinery and equipment, a fact which of course is of interest in a study of machine tools (World Bank, 1979, p. 4).

With the growth in industry, a structural transformation took place away from the agriculturally related industrial branches to intermediate goods and to the engineering sector (Sourrouille and Lucangeli, 1980). In Table 6.1, we can see the value added, in constant 1960 pesos, for both the manufacturing and the engineering industry. Already by 1965, the engineering industry had reached a 30 per cent share in value added, which is comparable to a developed country. The engineering industry also achieved a fairly high growth rate – a cumulative 8 per cent per annum between 1950 and 1974. Since 1974, however, the industry has stagnated, as has the manufacturing sector as a whole.

Table 6.1 *Growth in value added in the manufacturing industry and the share of the engineering sector in Argentina, selected years 1950–1979*

Year	Engineering sector (constant 1960 pesos m.)	Manufacturing industry (constant 1960 pesos m.)	Share of engineering sector in manufacturing industry %
1950	305	1,924	15.9
1960	810	2,878	28.2
1965	1,176	3,882	30.3
1970	1,549	4,978	31.1
1974	2,255	6,572	34.3
1975	2,083	6,388	32.6
1976	1,994	6,087	32.7
1977	2,216	6,314	35.1
1978	1,902	5,812	32.7
1979	2,236	6,451	34.7

Source: 1950–1976: World Bank (1979), Vol. 1, table 1 in the Statistical Appendix. 1977–1979: Banco Central de la Republica Argentina (no date).

The structure of the engineering industry is worth noting as it resembles that of a developed country to some extent, with the non-electrical machinery having an important share in the value added (see Table 6.2). The Argentinian engineering industry is, however, over-represented in the transport industry compared to some developed countries.

Table 6.2 *The structure of value added in the engineering industries of Argentina, the FRG, Sweden, the UK and the USA (%)*

Sector (3-digit ISIC)	Argentina (1973)	FRG (1978)	Sweden (1979)	UK (1979)	USA (1979)
Metal Products (381)	23.1	12.9	18.2	19.1	15.8
Non-Electrical Machinery (382)	20.0	29.1	29.5	28.9	29.5
Electrical Machinery (383)	16.1	25.4	18.8	19.5	20.7
Transport Equipment (384)	38.7	27.3	31.2	27.6	26.5
Scientific and Professional Instruments (385)	2.1	5.2	2.3	4.8	7.5

Sources: FRG, Sweden, UK and USA: United Nations (1980b). Argentina: Sourrouille and Lucangeli (1980).

Until the second half of the 1970s, the industrialization of Argentina took place within the framework of an import substitution policy where local production was stimulated by a number of policies, such as tariff protection and import quotas. Import substitution also took place in the engineering industry. In Table 6.3 we can see the domestic content of investment in machinery, equipment and transport equipment in Argentina from 1971 to 1979. Already in 1971 the domestic content was 75 per cent, a figure that rose to a peak of 85 per cent in 1973.

The domestic supply content appears to be extremely high for a country the size of Argentina. Measured somewhat differently, we can compare the pattern in Argentina with that of other small countries. For the engineering industry, the export ratio of production in Argentina in 1973 was only 5.2 per cent and the import ratio of investment only 12 per cent (World Bank, 1979, p. 113). Both these ratios are substantially lower than those of Sweden and Holland, whose output of engineering products and domestic market for such products is far larger than in Argentina. The figures for these two countries in 1979 were 45 and 48 per cent respectively for export ratios and 37 and 54 per cent respectively for import ratios of investment (Edquist and Jacobsson, 1982, pp. 21, 22).

6.2 The Argentinian machine tool industry

Whilst machine tools have been produced in Argentina since the beginning of the century, it was not until about 1960 that production began to take place on a larger scale. The increase in the production of machine tools in Argentina in the 1960s was intimately linked to the growth and diversification of the engineering industry discussed above (Castano *et al.*, 1981, p. 15). As Amadeo *et al.* explain:

Table 6.3 *Investment in machinery, equipment and transport equipment and the domestic content of investment in Argentina, 1971–1979*

Year	Total investment (1960 pesos m.)	Domestic supply (1960 pesos m.)	Domestic content of total investment %
1971	2,136	1,612	75.5
1972	2,244	1,744	77.7
1973	2,314	1,950	84.3
1974	2,392	2,028	84.8
1975	2,189	1,793	81.9
1976	2,240	1,891	84.4
1977	2,789	2,158	77.4
1978	2,223	1,661	74.7
1979	2,730	2,033	74.4

Source: Elaboration on Banco Central (no date).

This strengthening [of the machine tool industry] became finally effective at the beginning of the 1960's, when due to the need to rely on a real machine tool industry, capable of offering the elements necessary for the new model of industrial development that was being prepared for the country, attention began to be paid to the fact that in order to ensure a firm development of the sector it was necessary to grant sufficient custom protection. [Amadeo *et al.*, 1982, p. 5]

It is also significant that in a study made in the mid-1970s, 70 per cent of the machine tool firms dated their origin to the mid or end of the 1950s (Amadeo *et al.*, 1982).

As mentioned in the quotation above, the local machine tool industry was fostered through the erection of tariff barriers. According to one source, the nominal tariff rate for machine tools produced in Argentina was 175 per cent in 1965, which was modified to 80 per cent in 1971 and 50 per cent in 1978 (Ministerio de Economia, 1979).

Data on production and trade in machine tools for the period 1962–83 are shown in Table 6.4. Production rose from 5.4 million USD in 1962 to 62 million USD in 1979, growing by a factor of 11.5. However, production dropped dramatically after 1979. Exports were low for most of this period, fluctuating around 15 per cent of the value of output. Some increase in the export ratio can however be seen in the second half of the 1970s when exports were about 20 per cent of output. After 1979, the export ratio rose but to a large extent this rise was due to the profound reduction in the value of output. The import share of apparent consumption has always been high and has fluctuated around 50 per cent since the mid 1960s. From 1980, however, the import ratio rose dramatically, the reasons for which will be discussed below.

THE CASE OF ARGENTINA

Table 6.4 *Production and trade in machine tools in Argentina, 1962–1983*

Year	Production (current US$ m.)	Exports (current US$ m.)	Imports (current US$ m.)	Apparent consumption[a] (current US$ m.)	Exports/ production %	Imports/ apparent consumption %
1962	5.4	0.4	20.0	25.0	7.4	80.0
1963	5.5	2.0	11.1	14.6	36.4	76.0
1964	9.7	1.9	8.6	16.4	19.6	52.4
1965	16.9	1.5	6.0	21.4	8.9	28.0
1966	12.0	1.1	13.1	24.0	9.2	54.6
1967	7.9	1.1	11.3	18.1	13.9	62.4
1968	11.9	1.6	10.5	20.8	13.4	50.5
1969	17.6	1.7	16.6	32.5	9.7	51.1
1970	21.7	2.6	28.7	47.8	12.0	60.0
1971	22.2	3.9	24.1	42.4	17.6	56.8
1972	28.7	2.0	25.2	51.9	7.0	48.6
1973	38.3	4.7	23.1	56.7	12.3	40.1
1974	n.a.	6.7	26.9	n.a.	n.a.	n.a.
1975	42.0	9.5	29.6	62.1	22.6	47.7
1976	42.0	8.2	26.5	60.3	19.5	43.9
1977	60.0	8.5	37.0	88.5	14.2	41.8
1978	60.0	10.0	68.6	118.6	16.7	57.8
1979	62.0	19.2	58.6	101.4	30.1	57.8
1980	50.0[b]	27.7	95.4	117.7	55.4	81.0
1981	35.0[b]	20.0	70.0	85.0	57.1	82.3
1982	n.a.	n.a.	n.a.	n.a.	n.a.	n.a.
1983	28.2	14.0	23.5	37.7	49.6	62.3

Sources: Production data, 1961–73: Amadeo *et al.* (1982); 1975–83: *American Machinist* (various).

Export and import data, 1961–73: Amadeo *et al.* (1982), 1974–76: Ministerio de Economia (1979); 1977–79: data supplied by INDEC; 1980–81: *American Machinist* (various); 1983: NMTBA (1985/86).

[a] Production minus exports plus imports.
[b] Preliminary data.

The imports have consisted of high-priced machine tools, whilst the exports have been low-priced machine tools. The price per unit of traded machine tools is shown in Table 6.5. Until the mid-1970s the price per imported machine tool was often about ten times higher than the price of exported machine tools. After 1975, a change appears to have taken place, but one should note that from 1975 the data were taken from a different source. In any case, the picture still remains of Argentina as an exporter of machine tools of low value. Total exports have, however, been low; furthermore, the sole destination has been other Latin American countries, often within the LAFTA area. The destination of exports of machine tools in 1976 is shown in Table 6.6.

Table 6.5 *Price per unit of machine tools traded in Argentina, 1962–1978*

Year	Import price (US$ '000)	Export price (US$ '000)	Ratio %	Export price as % of import price
1962	10.5	1.8	5.8	17.1
1963	15.8	1.3	12.2	8.2
1964	12.3	1.0	12.3	8.1
1965	8.5	1.2	7.1	14.1
1966	16.3	0.8	20.3	4.9
1967	10.3	1.1	9.4	10.7
1968	9.5	1.3	7.3	13.7
1969	9.7	1.0	9.7	10.3
1970	9.9	1.0	9.9	10.1
1971	10.9	2.8	3.9	25.7
1972	25.2	1.2	21.0	4.8
1973	23.1	2.2	10.5	9.5
1974	n.a.	n.a.	n.a.	—
1975	15.4	3.9	3.9	25.3
1976	30.8	8.9	3.5	28.9
1977	41.4	10.1	4.1	24.4
1978[a]	23.3	4.7	4.9	20.2

Sources: 1962–73: Amadeo *et al.* (1982). 1975–78: CAFMHA (1979).
[a] January–September only.

Table 6.6 *Destination of exports of machine tools from Argentina, 1976*

Country	Value (US$ '000)	%
Mexico	3,152	38.6
Brazil	1,394	17.1
Bolivia	803	9.8
Uruguay	661	8.1
Venezuela	600	7.3
Peru	559	6.8
Chile	549	6.7
Paraguay	70	0.9
Ecuador	59	0.7
Other	322	4.0
Total	8,169	100.0

Source: Ministerio de Economia (1979).

6.3 Government policy

The nominal rates of protection for machine tools were given in section 6.2. For lathes, which are of more interest to us, the nominal rate was 150 per cent in 1960, but it gradually declined to 65 per cent in 1976. Some variations existed between the different types of lathes, engine lathes receiving the highest protection. Effective protection may, however, differ from nominal protection. According to one source, the effective protection for the machinery industry was 120.3 per cent in 1969 according to the Balassa method and 108.4 per cent according to the Corden method, while the nominal protection was 89.7 per cent (World Bank, 1979, p. 13).

Machine tools of the simple type produced in Argentina do not use a large amount of imported components. They do, however, use a large amount of domestically produced raw materials, e.g. castings and steel bars, whose price in relation to the international price affects the effective protection of machine tool firms. According to several sources, the price of important raw materials, such as sheet metal and steel, can be up to 100 per cent more than the international price (World Bank, 1979, p. 63; Association of Argentinian Engineering Industries, 1980). This pricing can be a very serious impediment to exports, as around 25 per cent of the cost of a simple machine tool can be accounted for by raw materials (see Table 6.7 and Table 4.1). A higher price for raw materials also means that the effective protection will be lower than the nominal protection. Let us assume that the international market price of a machine tool is 100 and that the value added is 63, as in Table 6.7. Assume as well that a nominal tariff of 65 per cent is introduced in Argentina and that it raises the local price by 65 per cent. Finally, assume that the prices of raw materials are not higher than the

Table 6.7 *Structure of costs of metalcutting machine tools in Argentina in the early 1970s (%)*

Cost item	%
Labour	30
Materials	37
of which:	
Casting	55
Sheet metal	15
Iron plate	3
Other	27
General costs	33

Source: DGFM (1972).

international prices. The effective tariff would then be 10 per cent.[1] If, on the other hand, all raw materials are priced 100 per cent higher than the international price, the effective tariff becomes 44 per cent.[2]

Since the nominal tariff was lowered to 65 per cent in 1976, the effective protection could well have been higher than 44 per cent prior to 1976. According to one study, however, the domestic price in 1971 of an engine lathe was only 59 per cent higher than the international price, even though a 90 per cent nominal tariff was granted. Hence, it appears that the full effect of the high nominal tariffs was below the potential. In spite of this, the effective tariff was high, possibly around 36 per cent in the early 1970s. (This assumes that the domestic price was about 60 per cent higher than the international price and that raw materials were priced 100 per cent above the international price.) The degree of protection received in the mid-1970s was also, according to one study, very much to the satisfaction of the machine tool producers. 'All the entrepreneurs interviewed stated that they felt adequately protected by the custom system in force' (Amadeo et al., 1982, p. 59).

In addition to the tariff system there are taxes that increase the real protection. There are, for example, what the World Bank (1979, p. 15) calls statistical and consular taxes as well as a tax on shipping costs. Together these taxes may amount to 12–15 per cent of the c.i.f. value of imported goods. A representative from a leading machine tool firm also emphasized this protection (interview).

According to the World Bank (1979, p. 23), there was a clear anti-export bias in the overall incentive system in Argentina at the end of the 1960s. In the early 1970s, the government strengthened the tax reimbursement scheme for exporters of capital goods, who received 35 per cent of the f.o.b. value of exports, a sum that was reduced to 25 per cent in 1977 but raised again, at least for machine tools, to 35 per cent in 1981 (interview with machine tool builder; World Bank, 1979). A number of other export incentives were also introduced, but according to one source the reimbursement constituted the greatest part of the incentive (World Bank, 1979). In the case of machine tools, this observation is strengthened by the fact that in a study in the mid-1970s, whilst entrepreneurs were satisfied with the export promotion policy in terms of the automatic reimbursement scheme, none of the twenty-two firms interviewed used other programmes set up to promote exports (Amadeo et al., 1982, p. 59). Hence, it would seem reasonable to suggest that, at least in the case of machine tools, the export promotion schemes did not fully change the bias towards the home market that the high tariffs had created.

In summary: I have suggested that the effective protection prior to 1977 was high, even though it was lower than the nominal protection on account of the high prices for domestically produced raw materials.

THE CASE OF ARGENTINA 121

A very generous automatic reimbursement scheme was implemented in the 1970s in order to promote exports. However, only about 20 per cent of the value of the production of machine tools was exported in the second half of the 1970s. One possible explanation of this weak response to the export incentives could be that it was more profitable to sell on the local market than abroad because of the total structure of incentives. It is probably true that there was a greater incentive to sell on the local market, but that in itself does not explain the very low export ratio.

Given the limited local market, why did the firms not grow by utilizing the generous reimbursement scheme? After all, with a small domestic market, there is really no reason why a structure of incentives favouring the local market would drastically limit exports if exports were not directly penalized. In trying to answer this question, we need to go beyond the traditional *structure of incentive analysis* and introduce into the analysis the risks of pursuing export-oriented growth in a context of rapidly fluctuating relative prices. As was mentioned in section 4.4, exporting machine tools (including conventional ones) involves a substantial amount of marketing investment in order not only to build up a marketing network (which can be leased) but also to create a brand name. Either the machine tool producer or the distributor, if one is used, needs to invest heavily in order to break into the market. An export orientation thus requires a fairly long-term view on strategy issues, which in turn requires some degree of certainty about the price competitiveness of the exported goods.

A fundamental characteristic of the Argentine real exchange rate is that it fluctuates enormously. The yearly fluctuations between 1968 and 1979 are shown in Table 6.8. Analysed by quarters it can also be mentioned that in 25 out of 45 quarters in this time period there was a change of more than 5 per cent in the real exchange rate and in 10 out of 45 quarters the change was greater than 10 per cent. An export-oriented strategy under such conditions would involve very high risks because it would of course be impossible to plan a product's price competitiveness in a given foreign market even with a one-year planning horizon. This particular point was strongly expressed by firm representatives. This could then explain the common statement that Argentinian producers only export when there is low domestic demand (Valeiras *et al.*, 1978, p. 35, for example).

6.4 The diffusion of CNC lathes in Argentina

In section 3.3.1, I argued that the very large and captive domestic market for CNC lathes in Japan reduced the risks associated with the implementation of the overall cost leadership strategy by some Japanese firms. The existence of a relatively large domestic market (in

Table 6.8 *Fluctuations in the real exchange rate[a] in Argentina, 1968–1979*

Year	% change	Year	% change
1968	− 2.8	1973	− 6.3
1969	− 3.9	1974	+32.3
1970	−10.7	1975	− 7.0
1971	+31.4	1976	+14.4
1972	−26.3	1977	−22.2
1973		1978	−41.9
		1979	

Source: Yuravlivker (1980).

[a] Defined as $\dfrac{E}{PA} \times PUS$

where E = exchange rate
PA = price index in Argentina
PUS = price index in the USA

relation to the minimum efficient scale of production) is, of course, also important for firms in the NICs. A significant factor in the choice of strategy for firms would then be the actual and, especially, the potential domestic market for CNC lathes.

In the case of Argentina, a leading North American CNC lathe producer claimed in 1981 that it would be able to sell 100 CNC lathes per annum during the following few years (interview). The total market must then be greater than this as it is unlikely that one firm can supply the whole market. The Instituto Nacional de Technologia Industrial (INTI, 1981), in a study in collaboration with the German Ministry of Science and Technology, suggested that the annual market for CNC units would be 990 over the following few years. As around 40 per cent of the installed CNC machine tools are usually CNC lathes, a market of around 400 units per annum is thus envisaged.

In this section, I shall suggest that these estimates are far too high. First, however, I shall present an analysis of the historical rate and pattern of diffusion of CNC lathes in Argentina; only then shall I attempt to make a more realistic estimate of the potential demand for CNC lathes.

6.4.1 *The historical rate of diffusion of CNC lathes in Argentina*
The results of a survey of the historical rate of diffusion of CNC lathes in Argentina are presented in Table 6.9. By April 1981, a total of 189 CNC lathes had been identified.[3] It is reasonable to add another 15 per

cent to this figure as some units, particularly older ones, were certainly overlooked. Hence, I estimate that around 215 CNC lathes were installed in Argentina by the beginning of 1981. For the period after April 1981, data on production and import are available from Center on Transnational Economy (1984). According to that source, total production and imports were 45 units in 1981, 9 units in 1982 and 11 units in the first eleven months of 1983. However, the data, for 1982–83 at least, probably underestimate the flow. For example, in 1983 imports are said to be 1 unit, but Japan alone exported eleven CNC lathes to Argentina in 1983. In addition, exports from the EEC countries amounted to approximately 2 million USD in 1983, which could be equal to 10–15 units. I would therefore suggest that the stock approached 300 units at the end of 1983.

Table 6.9 *The instalment year[a] of the known CNC lathes in Argentina, 1976–1981 (units)*

Year	Number
1976	5
1977	17
1978	30
1979	33
1980	45
1981[b]	21

Source: Survey by the author.
[a] Instalment year known in 151 cases (80% of total).
[b] Only to May.

Apart from the overall improvement in the technical and economic aspects of CNC lathes in this period discussed in Chapter 2, two further factors are worth mentioning with regard to the process of diffusion in Argentina.

(i) Availability of the skills necessary to use CNC lathes Although skill-saving in nature, CNC lathes need some specific skills in order to be diffused. There are four types of skills that are absolutely essential for a diffusion of this new technology. These are: maintenance of both the electronic and mechanical parts, programming skills, skills to realize the need for CNC lathes, and skills to utilize CNC lathes.

The CNC systems are usually maintained by the local branch of the electronic firm that produced it, e.g. General Electric or Siemens, and not by the supplier of the machine tools, which may be a local firm. An electronic firm needs 15–30 units installed in an economy in order to justify employing a qualified engineer for maintenance. With a slowly growing market for NCMTs, and with a multitude of brands, problems of a shortage of this type of skill must exist and certainly have arisen in Argentina. Of course, with a faster-growing economy, the CNC supplier's expectations may be such that he takes the initial costs of overheads.

I noted earlier that the mechanical parts of the machine tool have been altered significantly in response to the opportunities offered by CNC. This has also been a source of maintenance problems. Thus, a similar skill shortage exists for the mechanical parts, even if it is not so accentuated as for the electronic parts.

One may also note that the division of labour between the users and suppliers of CNC units has altered very little over time. Most importantly, this means that the *teaching* function of the supplier is still extremely important, in relation both to programming, where progress is taking place, and, most importantly, to the utilization of the new machinery. For example, production planning, tool choice, etc. may alter significantly with the use of CNC lathes.

Therefore, the very slow rate of diffusion in Argentina could partly be explained by a deficient supply of all the above types of skills prior to 1977–78. In these years two more CNC firms started up in Argentina, which could partly explain the jump in the rate of diffusion for these years. However, it is also important to note that the establishment of a maintenance function, not always covering teaching, by these firms was in *response* to the demand from their customers who had seen the brands in question in exhibitions in Europe. Thus, there has been and may still exist a lack of will to invest in the rather expensive overheads related to the service and maintenance functions. Interviews with firms in the metalworking sector strongly point to the importance of an adequate supply of these skills. For example, one prospective buyer said when asked what his criteria were for the choice of CNC lathe: 'The one that supplies the best service and maintenance on the electronic part' (interview).

(ii) **Public investment** The market for Argentina's metalworking sector has been constant since 1974. Thus, at an aggregate level, there has been very little stimulus for investment in new equipment in the Argentinian economy. However, whilst the private sector has reduced its investment since 1974, public sector investment has become much more dynamic and grew by 67 per cent between 1974 and 1978 (Calvar *et al.*, 1980). In particular, investments in the petroleum, hydro and atomic power sectors have generated a demand for CNC lathes among the suppliers of parts and components, e.g. valves and pumps. Indeed, about 50 per cent of the stock of CNC lathes in 1981 were found in firms that directly serve sectors where there is much public investment (survey by the author).

6.4.2 An estimate of the future rate of diffusion

The size of the market for CNC lathes in an economy is a function of both the economic profitability of the technology in relation to conventional machine tools and the number of times this choice is made. The latter depends on the size of the industrial branches that use lathes in their production process and the rate of investment in these industries.

It was argued in Chapter 2 that CNC lathes reflect a technical change that may well be as advantageous to investors in developing countries as to investors in developed countries. The critical question was the scarcity of skilled labour. However, even if skilled labour was more abundant in the developing countries, the break-even wage for choosing CNC lathes was shown to be fairly low. This being the case, the size of the market for CNC lathes should then be *mainly* a function of the size of the relevant industrial branches and the rate of investment in these.

To understand the differences in the use of CNC lathes by branch, the diffusion pattern in Sweden was studied at the four- and five-digit ISIC level. The data refer to 1976. It was found that eight subsectors contained 85 per cent of the CNC lathes. These are: Tools and Implements (3811); Engines & Turbines (3821); Agricultural Machinery (3822); Metal and Wood Working Machinery (3823); Special Industrial Machinery (3824); Other Non-Electrical Machinery (3829); Electrical Machinery (3831); and Automobile Parts (38432). Hence, it is the size and growth pattern of these eight branches that are of interest to us in determining the relationship between industrial structure and the market for CNC lathes.

In the Argentinian case, data from INDEC[4] show that, using a moving average, the importance of these eight branches grew from 26.7 per cent to 37.6 per cent of the value of production in the engineering sector during the 1970s. The data also show that three of the fastest-growing branches were Special Industrial Machinery (3824), Other Non-Electrical Machinery (3829) and Automobile Parts (38432). In Sweden in 1976 these three branches contained 61 per cent of the CNC lathes, and also showed the highest intensity in use of CNC lathes. Hence, there has been a change in the structure of the Argentinian engineering sector that appears to be conducive to the demand for CNC lathes. We still have to keep in mind, though, the relatively small absolute level of production in Argentina. It was estimated that the value of output in the eight relevant branches amounted to only 2.5 billion USD in 1977/78 whereas in Sweden the equivalent figure was over 7 billion.

The method used to estimate the potential demand for CNC lathes is a simple one. It is assumed that the variations in intensity in use of CNC lathes between branches reflect their different technical characteristics of production. For example, Special Industrial Machinery, which has one of the highest intensities in use of CNC lathes, has more metalcutting than the production of Office Machinery. I further assume that the content of production at the four- and five-digit levels is roughly the same in Sweden and Argentina.

After calculating the estimated sales value in Argentina in 1977–78, I multiplied this figure at the four- and five-digit ISIC level with the intensity in use of CNC lathes in Sweden in 1976. I measured intensity by the ratio between the number of CNC lathes installed at the four- and five-digit level and the sales value at the same level of aggregation. This '1976 potential' can be seen in Table 6.10. The eight branches in Table 6.10 accounted for 85 per cent of the CNC lathes installed in Sweden in 1976. We thus need to add another 35 units to the '1976 potential'. The total potential would then be 234 units, which was very close to the actual stock in 1981 (April). Hence, one could suggest that there was a five-year lag in the use of CNC lathes between Sweden and Argentina.

Table 6.10 *The '1976 potential' for CNC lathes in Argentina (units)*

Branch	'1976 potential'	Actual stock (early 1981)[b]
3811	3	31
3821	6[a]	7
3822	28	1
3823/4	54	23
3829	23	46
3831	13	1
38432	72	43
Total	199	152

Sources: survey by the author and elaboration of Steen (1976/1977).
[a] UK data.
[b] I have excluded the CNC lathes for which I have no information on branch distribution.

In Sweden, though, the stock of numerically controlled machine tools doubled between 1976 and 1981. So, if we assume that the proportion of CNC lathes to all numerically controlled machine tools was the same in 1981 as in 1976 (40 per cent), we need to double the 1976 potential in order to get a correct estimate of the '1981 potential' for Argentina. This potential would then be 468 CNC lathes.

In other words, in order to reach the same intensity in use of CNC lathes as in Sweden in 1981, the total number of CNC lathes would need to be 468. In this exercise I have used the level of sales in 1977–78 in Argentina as the denominator. If the economy expands, we would need to add a given number of new CNC lathes to the potential. Table

6.11 shows one way of estimating the demand for CNC lathes in the period 1981–85. I differentiate between CNC lathes sold to realize the given potential and new potential as a result of an expansion of the sales in these eight branches. I assume in this example that the realization potential is reached in five years and that there is a continuation of the historical growth rate of 5.9 per cent in the eight branches concerned.

Table 6.11 *An estimation of yearly demand for CNC lathes in Argentina, 1981–1985 (units)*

Year	Realization potential[a]	Extra potential	Total
1981	51	21	72
1982	51	25	76
1983	51	29	80
1984	51	33	84
1985	51	37	88

[a] $(468-215) \div 5$.

The limitations of such a mechanistic analysis are many. First of all we need to remember the assumptions. Secondly, the peculiarities of the Argentinian situation should be borne in mind. For example, the economy is very erratic, as well as being a very small market. There is therefore a large premium on flexibility of the production equipment. In practical terms, this means that in some instances in Argentina CNC lathes would be chosen instead of an automatic, less flexible lathe, whereas that would not have happened in Sweden. Thirdly, technical change continuously widens the potential for CNC lathes. For example, the cost of programming, which is a fixed cost, is decreasing all the time. This means that the minimum size of batch for which CNC lathes are efficient is constantly being reduced. However, in spite of the many limitations, the *magnitude* of the size of the domestic market would be closer to 100 CNC lathes than to the estimates made by INTI and the foreign supplier mentioned at the beginning of this section.

Whether the potential will be *realized* is another matter, which primarily depends on (a) the availability of the skills required to maintain and service the CNC lathes as well as knowledge of CNC technology among the firms in Argentina, and (b) the rate of investment. Concerning the first factor, it seems probable that the present growth rate of installed CNC lathes will help to establish a service and maintenance structure. This is evidenced by the large number of smaller firms that have bought one CNC lathe (see below). Furthermore, it will diffuse information about CNC lathes and allow for an accumulation of experience in using CNC lathes.

As far as the rate of investment is concerned, the Argentinian economy has been in an exceptionally severe crisis for years. For example, in 1980 the level of capacity utilization in the machine tool industry was close to a mere 30 per cent while several capital goods industries had a capacity utilization of less than 50 per cent (Association of Argentinian Engineering Industries, 1980). Another example is that production in the capital goods sector was 25 per cent lower in 1983 than in 1974. Finally, the level of investment in 1983 was only 52 per cent of the level in 1977 (Center on Transnational Economy, 1984). Of course, if the potential market for CNC lathes is to be realized, the investment level will need to increase.

6.4.3. Conclusion

The overall conclusion is that the domestic market for CNC lathes in Argentina is, and will stay, much smaller than that envisaged by both individual CNC lathe producers and INTI.

From the point of view of the lathe producer in Argentina (analysed in section 6.5) a further factor must be taken into consideration, namely, that there is a segmentation of the market for CNC lathes in the Argentinian economy too. Around 40 per cent of the CNC lathes installed are high-performance CNC lathes, which are, as a rule, sold to firms catering for the state market or to foreign firms in the automobile industry. The former market is, however, the dominant one. The CNC lathes of medium and low performance are predominantly of Italian origin. These lathes tend to be sold to small subcontractors and to small machinery and component producers, which often have only one CNC lathe. Indeed, only about 25 per cent of the stock of CNC lathes were installed in firms having more than 6 units (survey by the author). For example, only 8 per cent of the leading Italian firm's CNC lathes were sold to (the often large) firms catering for the state demand.

The segmentation of the market means that the market available to the local firm, which cannot, and does not, pursue a strategy that involves selling to the segment demanding high-performance CNC lathes, is around 40 per cent smaller than the total market. Furthermore, the firm would need to compete with some very successful Italian firms for the remaining, say, maximum of 50 units per year.

6.5 The firm producing CNC lathes

In the mid-1970s, twenty firms were producing lathes in Argentina. The total production of lathes in Argentina fluctuated for a long time around 2,000 units annually, as can be seen in Table 6.12. It seems likely that the number has decreased in recent years in view of the drastic decline in machine tool production in Argentina after 1979. The three largest firms accounted for 45 per cent of the production of lathes in units, according to industrial sources, and far more in terms of value (Association of Argentinian Engineering Industries, 1980). The trade pattern follows that of machine tools generally, with low exports – in 1975, 23 per cent of the units produced were exported – although the share of exports appears to have grown substantially in the past few years and, according to some data, the share in terms of units was as high as 50 per cent in 1979 (CAFMHA, 1979). Imports are, however, far higher than exports, if we measure in terms of value. In 1975, imports amounted to 6.4 million USD while exports reached 1.5 million. In 1979, the sums were 16.7 million and 5.2 million respectively (CAFMHA, 1979). Price per unit differs greatly between imported and exported units, the latter being worth only about one-tenth of the former (see Table 6.13).

Table 6.12 *Production of lathes in Argentina, 1957–1979 (units)*

Year	Units	Year	Units
1957	2,884	1969	1,552
1958	3,049	1970	1,853
1959	2,972	1971	2,003
1960	3,720	1972	1,663
1961	3,580	1973	1,973
1962	2,082	1974	2,019
1963	1,204	1975	1,623
1964	1,550	1976	2,895
1965	2,091	1977	n.a.
1966	1,576	1978	2,660
1967	993	1979	2,361
1968	1,325		

Sources: 1957–76: Castano et al. (1981). 1978–79: data received from the Association of Argentinian Engineering Industries.

Table 6.13 *Price per unit of lathes exported and imported in Argentina, 1975–1978 (US$)*

Year	Export price	Import price	Export price as % of import price
1975	4,130	27,729	14.9
1976	4,585	62,859	7.3
1977	4,665	45,730	10.2
1978	3,481	40,083	8.7

Sources: Imports 1975–78 and exports 1975–77: CAFMHA (1979). Exports 1978: data received from INDEC.

Thus, the lathe industry can be characterized as consisting largely of small units producing simple lathes mainly for the domestic market, even though some change in the orientation of sales seems to have been taking place lately. Three out of the twenty firms have risen above the average in terms of size.

Only one of the twenty firms produces CNC lathes. On the basis of Castano *et al.* (1981), I shall briefly describe this firm's history. It was established as early as 1937 when two skilled workers of Italian origin left a larger engineering firm to set up a small shop for the production and repair of machine tools. By 1953, annual production was around 100 lathes and twenty workers were employed. The firm was then in a medium position within the industry. In 1959, the firm was acquired by a foreign group. Following the change in ownership, drastic changes were made in the organization of the firm and in the types of products produced. Large investments in new machinery were also made. The firm was transformed from a production unit of an artisanal nature into a modern factory.

The transformation of the firm coincided with the growth of the engineering industry in Argentina. The policy of fostering the engineering sector created a demand for high-quality machine tools, which hitherto had not been produced on a large scale in the country. Only a few of the leading machine tool firms were able to respond to this new type of demand; most firms continued to cater for the highly priced elastic market of repair shops, etc.

The firm's policy of investing in high-quality equipment as well as of strengthening its design team enabled it to expand its share of the market rapidly from 7 per cent in 1960 to 38 per cent in 1963. The firm has since then been the market leader in Argentina, although its share of the market has fluctuated somewhat: in terms of value, its market share varied between 30 and 49 per cent for the years 1966–1976.

Since the early 1960s, when the firm initiated its first prototype development, great emphasis has been placed on product design in its long-term strategy. The strategy of the firm has been to utilize these superior (in the local context) design skills to cater for the high-performance segment of the market. The strategy has also involved a widening of the product mix to exploit small niches of the domestic market for more advanced lathes. The output of the firm has thus included not only engine lathes but also revolver lathes, semi-automatic lathes and a milling machine, as well as a large number of variations in the size of each of the lathes. The firm has mainly sold its machines on the local market, exceptions being the export of a number of copying lathes to Brazil under the LAFTA trade agreement. Given the structure of incentives discussed above, the inward orientation of the firm is not surprising. Given the reliance on the local market, and its strongly, cyclical nature, the firm's niche mentality can be seen as a way of reducing the risks still further.

At the end of the 1970s, the firm decided to introduce CNC lathes into its product mix and in 1980 the first batch was produced.

In section 5.1 I analysed the strategic position of the Agentinian firm within the international industry producing CNC lathes as well as within the group of firms pursuing the low-performance strategy. The conclusion was clearly reached there that the Argentinian firm has characteristics that put it in the group of firms in the worst position in the world industry. I shall elaborate more upon these characteristics.

First, the firm is *small* internationally speaking, with only 150–200 employees and a sales value of around 5 million USD. This means that the firm has a weak financial basis. It also means that, although 4 per cent of the staff are design engineers, the total number of engineers performing this function is very small, less than 10 people. In addition, the design team was headed by an *inventive mechanical engineer* rather than being a proper team. What is more, the team included only one electronic engineer.

Secondly, the firm has a very *diversified* output. A high degree of diversification is normally serious when a firm is very small because major problems arise in the planning of production. Diversification also leads to an excessive generation of production planning skills: 7–8 per cent of the staff were production planners in this firm, whereas the equivalent number was only 3.6 per cent and 2.9 per cent respectively for the leading Taiwanese and Korean firms. The diversified output also meant that it was not possible to concentrate design efforts on one or two types of machine tools. This is reflected in the long time (four years) that it took to design its CNC lathe. The major Asian NIC producers take less than half this time, a fact of importance given the rapid rate of technical change in the industry (see section 4.1).

Thirdly, the *lack of exports* to countries other than Brazil and Mexico means that the firm has no developed marketing network in the USA or in Europe. The lack of exports also seems to have resulted in a lack of information on the technological frontier. It is clear that the design concept was already dated even when the first CNC lathe left the factory. Finally, the inward-looking strategy led to a situation where the management was not well acquainted with the outside world, and did not emphasize the creation of skills required to analyse the conditions for success within the international market.

In summary, the firm is weak in three critical aspects: size and associated financial strength; skills; and marketing network.

I shall now compare the resources of the firm with the requirements determined by the nature of competition in the industry. This will be done by discussing the two strategies that are open to the firm: the first involves expansion on the world market along the lines discussed in Chapter 5, and the second involves a more limited production of CNC lathes aimed primarily at the regional market, but behind protective barriers.

6.5.1 The full export-oriented strategy

This strategy implies first improving the firm's position within the low-performance strategy, i.e. moving northeast in Figure 5.1, and eventually shifting to the overall cost leadership strategy. Three basic conditions for a successful pursuit of this strategy for this particular firm will be discussed: the scale of output must be increased in conjunction with a specialization in CNC lathes; a competitive pricing policy must be pursued; and design skills must be improved.

(i) Scale of operation and specialization In order to assess the break-even point for this particular firm, an analysis was made with a representative of the firm of the advantages of increasing the scale of output and concentrating production on CNC lathes alone. Table 6.14 gives an account of the estimated advantages from specialization and scale in the firm. The table suggests that, if production is increased from a few CNC lathes per month to about thirty per month, production costs could be reduced by 25 per cent per unit of output. A little more than a quarter of this reduction reflects the advantages from specialization.

Advantages from specialization come from a number of sources:

- As a CNC lathe contains fewer moving parts than a conventional lathe, the light machinery section in the firm could be reduced. This is particularly important in a developing country that does not have an adequate component industry. This lack of *back-up industry* has meant that the firm itself has had to produce a range of components that a European firm would have bought from a specialized producer. The result in Argentina has been low machine utilization and therefore high costs. Most of this section could be eliminated.
- The cost of some special equipment related to the large number of models could be reduced to half. The large number of production engineering staff who plan a very diversified production mix could be reduced.
- The number of workers could be reduced.

Table 6.14 *An estimation of advantages from specialization and scale in an Argentinian firm producing CNC lathes*

Source	Option		
	(1)	(2)	(3)
Components	100	100	85
Capital	100	82	51
Labour	100	76	60
Other	100	100	75
Total	100	93	75

(1) Mixed output including a few CNC lathes per month.
(2) Specialization in CNC lathes. The data reveal only specialization effects.
(3) Specialized production of 30 CNC lathes per month. The data include advantages from both specialization and scale.
Source: Elaboration of data supplied by the firm.

The sources of scale advantages are:

- discounts for components,
- fixed costs for buildings and inventories,
- increased machine utilization through shift work,
- fixed costs for design and development as well as for production engineering,
- fixed costs in administration, and
- some fixed costs for electricity, etc.

We can see their relative importance in Table 6.15. Discounts for materials is by far the most important source of scale advantages (53 per cent); it should be noted that only a 15 per cent discount was assumed to exist when thirty units are produced per month. This is reflected in the cost structure (estimated) with a production of thirty units per month (see Table 6.16). The second most important source of scale economies is fixed capital (30 per cent). Scale economies in the use of fixed capital are generally low in this industry but in this particular case the firm had previously over-invested in fixed capital. This meant that the capital cost per unit output would decline significantly as the output increased. Scale economies arising from indivisibilities in R&D and in production planning accounted for only 14 per cent of the total gains from scale.

Whilst a detailed account of prices is not possible for reasons of confidentiality, a production of thirty units per month was judged to be indispensable for market success. The cost/pricing exercise assumed, however, a 25 per cent export rebate on the f.o.b. price, which was the size of the reimbursement programme discussed in section 6.2 above.

Table 6.15 *Sources of scale advantages (estimated) with a production of 30 CNC lathes per month in an Argentinian firm*

Source	%
Components	53
Capital	30
White-collar workers	14
Other	3
Total	100

Source: As for Table 6.14.

Table 6.16 *Cost structure (estimated) with a production of 30 CNC lathes per month in an Argentinian firm*

Cost item	%
Components	72
Capital	13
Labour	13
Other	2
Total	100

Source: Elaboration of data supplied by the firm.

(ii]) Pricing policy As noted above, the firm's current skill and production structure reflects a strategy of diversification on the local market. In a climate of free trade, the firm would have to specialize production and, associated with this, change the structure of skills – for example, increase the number of skilled assembly workers. In the production of CNC lathes, assembly is relatively more important than metalcutting, whilst the opposite is the case in the production of engine lathes. The readjustment would take a minimum of two–three years because skilled assembly workers in this field do not exist in Argentina. Hence, it would take two–three years before the level of production could be increased to thirty units per month taking only the skilled labour aspects into consideration. Given the problem of creating a brand name and breaking into the market for CNC lathes, the time horizon would probably be much longer.

Meanwhile, the firm would have to compete with the Japanese firms and others. Furthermore, as discussed in section 4.4, a newcomer needs to lower its price in relation to more established firms in order to sell. This means that the CNC lathes produced by this firm would need to be priced near to, and probably below, the long-run average cost. This would entail large initial losses until scale economies could be reaped.

(iii) Skills The firm has designed two CNC lathes of the same basic concept. The lathes are of the *old* type discussed in section 5.1 and thus use very dated design concepts. The choice of design concept was the result partly of the relatively weak design skills of the firm, and partly of relative ignorance of the rapid changes taking place internationally in terms of the design of CNC lathes.

The firm currently has seven designers. The leading firms within the low-performance strategy employ five to nine times this number of design engineers, and the leading firms in the global industry have thirty-six times the number. Given the long gestation period involved in building up the core designers, as discussed in section 4.1, a very substantial skill generation with high initial costs would need to take place if the firm is to be able to compete internationally at some point in the future.

The fulfilment of these three basic conditions would mean a drastic change in the firm. Reaching the break-even point of thirty CNC lathes per month would mean a complete shift in the type of markets served by the firm. From traditionally being oriented towards the domestic market, the firm would need to export 80–90 per cent of its production of CNC lathes, the domestic market being limited to a maximum of ninety CNC lathes per annum and only a part of this market being open to this firm (since a large part of the domestic demand is for high-performance CNC lathes). The firm would therefore need to build up and finance a marketing network abroad with all the associated costs. Alternatively, it could use an already existing distributor, which also would involve high costs. It would also need a very large amount of capital to finance sales expansion by 300–400 per cent (in the order of 20 million USD). During the period of expansion, the firm would also need capital to finance a low-price policy discussed above. Finally, a considerable investment in skill formation would be required.

The very wide discrepancies between the resources of the firm and the requirements set by international competition meant, though, that this strategy is not open to the firm. One option would then be to collaborate with a foreign firm and, indeed, in 1980–81 the firm was discussing the possibilities of doing subcontracting work for a European firm. Another possibility would be to reconstruct the firm within the framework of an Argentinian industrial policy.

The reasons for collaborating with a foreign firm would be threefold: the needs for capital, technical information (design) and a marketing network. One possible strategy for the firm would then be, at least in the short term, to acquire a licence and with it some risk capital and access to a marketing network. In other words, it would involve a relocation of production to a peripheral country (seen from the point of view of the developed country firm) under the control of the foreign firm. The central element would be the licence. Let us elaborate on the possibility of such an agreement.

For a licensor, there are two reasons for being interested in selling a licence to the Argentinian firm. The first would be to get access to the Argentinian and the Brazilian markets. The Argentinian and Brazilian governments recently agreed on a Brazilian reduction in tariffs on CNC lathes from Argentina to 5 per cent; it is far higher for imports from other countries. As the Brazilian intensity in use of CNC lathes is far below that of Argentina (Jacobsson, 1982b), a low-cost CNC lathe imported from Argentina should have a substantial Brazilian market, apart from the more marginal local market. This could of course be a sufficient motive for a foreign firm to sell a licence. Indeed, Pontiggia, the largest Italian lathe producer, one of the few successful European competitors to the Japanese, has sold a licence to a Brazilian firm. However, the limited size of the local markets would seem insufficient to justify the relatively large input of capital that would be needed for the reorganization and running of the firm. Hence, a licence from a foreign firm would have to be combined with another source of finance.

The second reason for a foreign firm to be interested in collaboration would be that it believed that the Argentinian firm could produce at a lower cost than the licensor. There does seem to be a view in some firms that a localization of production in a peripheral country could have this effect. However, the data on the cost structure in Table 6.16 make it difficult to understand such a view. The only sources of comparative advantage that this firm would have would be (a) lower direct labour costs, which account for about 11 per cent of the production costs in an advanced country like Sweden (interview), (b) a skilled labour force that is willing to work shifts, and thus increase machine utilization, and (c) government incentives, e.g. the 25 per cent export rebate.

To base a long-term investment decision on these three factors would, however, seem very unlikely. First, technical change involving the development of various automatic feeding mechanisms and material handling systems is now allowing for a 24–hour run of the capital stock with the aid of very few direct workers. Secondly, the export rebate could easily be abolished. That leaves lower direct labour costs and a risky rebate as reasons for a large-scale capital investment. On the negative side there are all the risks of investing in a peripheral country, especially given the exchange rate policy of the government, as well as the high prices of domestically produced materials and the need to have larger stocks owing to uncertain deliveries of imported components. All in all, while the production costs would probably be slightly lower in Argentina than in an advanced country, the cost reduction would not justify the risks associated with an investment of the magnitude required. This becomes clearer when we realize that a decision to locate production in a peripheral country would be part of a strategy to combat the Japanese. The maximum cost reductions attainable in Argentina would not go very far in breaching the gap to the Japanese. In this particular branch, such a simple solution to the problems faced by the European and US producers does not exist. I would therefore suggest that the possibility of a foreign firm participating in financing a strategy based on exports to the industrialized world is very small. Indeed, the discussions with the European firm about some kind of collaboration broke down.

What about the possibility of reorganizing within the framework of an Argentinian industrial policy? The biggest issue for the firm was the availability of capital. The firm needed capital to finance:

- the transition to large-scale and specialized production,
- a pricing policy enabling it to gain market shares abroad,
- work in progress, which would be dramatically higher than before,
- a long-term improvement in its design skills.

This capital would be high-risk capital given the weak position of the firm within the international industry producing CNC lathes.

One could argue that government intervention to provide risk capital is justified. The main argument for intervention (see Chapter 9) rests, however, on the assumption that the state's discount rate is lower than the firm's, i.e. that the state has a longer-term view of technological development than the firm. *In this particular circumstance*, it is very difficult to claim that this is the case. Of course, one could argue that a government *should* have a long-term view of capability generation. However, at the time when the firm had to take a strategic decision (1981–82), the implicit discount rate revealed by the government's industrial policy was clearly higher than the firm's. Governmental policies had revealed such an exceptional bias against the local production of capital goods that, assuming reasonably informed officials, one could not but conclude that the state saw no value in establishing lasting technological capabilities.

Since 1976, the policy of several governments has been to reduce nominal and effective tariffs. By 1977, the average nominal tariff for machine tools was only 25 per cent and the effective tariff was 10 per cent (World Bank, 1979). Simultaneously, a policy of planned overvaluation of the peso began in 1979, only to be be discontinued in 1981 with several massive devaluations. The environment for local firms was further complicated by the existence of high real interest rates: 2.8

per cent in the second half of 1977, 2.1 per cent in 1978, 0.1 per cent in 1979, 4.9 per cent in 1980 and 12.4 per cent in the first quarter of 1981 (Canitrot, 1981, p. 145).

Thus, at the same time as the tariff rate was rapidly reduced, causing in itself large adjustment problems, the peso was overvalued. This overvaluation further increased the difficulties for local machine tool firms, as evidenced in the estimated 82 per cent import share of investment in machine tools in 1981 (see Table 6.4). Finally, readjustment was made very difficult because of high interest rates. One firm even talked about real interest rates of 30 per cent (interview)! All in all, the situation in 1981 for the two largest producers was disastrous, with 30 per cent capacity utilization, while other capital goods producers had only 50 per cent capacity utilization rates.

On the other hand, government continued with its predecessor's policy of paying a rebate on the f.o.b. price of exported machine tools. This rebate is essential for creating price competitiveness for the firm in export markets, in both the short and medium term. Some of the rebate can be seen as compensation for the high domestic costs of materials, such as steel, but the greater part of it is a pure export subsidy. The subsidy would, in the case of a production of thirty units per month, amount to an equivalent of half the local content of production costs (excluding profits of the firm). This is, of course, a very large subsidy. Most importantly, as there are large-scale advantages, the subsidy permits a pricing of the CNC lathes that is closer to the long-run average costs. This effect has, however, happened more by accident than by design.

To some extent, the export subsidy takes care of the problem of risk capital – that is, the capital associated with pursuing a marketing policy with a very long-term view on profits. However, it is conducive only to the *production* of CNC lathes and not to the local development of designs. Furthermore, it is not sufficient to solve the problem of working capital, which is bound to increase as production is confined to CNC lathes and expanded to reap the gains from scale economies. Most importantly, however, the uncertainties involved in trading owing to exchange rate fluctuations mean that, in spite of the subsidy, firms will not take long-term decisions on exports. A more active specific government policy would therefore be required for this firm. (A policy ensuring a stable real exchange rate would, of course, be necessary, apart from firm-specific policies.) See Chapter 9 for a discussion of how governments can design intervention policies.

The firm could not, however, count on an active and specific government policy of the Korean or Taiwanese types (see Chapters 7 and 8). In contrast, it operated in a very hostile environment caused by the economic policy of the government. Nor could it count on any large-scale collaboration with a foreign firm. This meant that the firm's option of an export-oriented growth strategy was not valid. Let us therefore turn to the second scenario, which involves a penetration of local markets alone.

6.5.2 The regional market strategy

As noted above, the Argentinian and Brazilian governments agreed recently to reduce to 5 per cent the tariffs on Argentinian NCMTs, amongst other products. This is very important because the Brazilian tariff on other countries' products is prohibitive and Brazil's engineering sector is three to four times the size of Argentina's. In the absence of any large-scale local producer in Brazil, the Brazilian market would seem to be very important for an Argentinian firm producing CNC lathes. The size of the market is, however, rather limited: a total stock of 200 CNC lathes existed in Brazil in 1979 (Stemmer, 1981). Even though one would expect the size of the market to increase with the introduction of a CNC lathe priced not far above the international price, the combined markets of Argentina and Brazil would be below the required minimum level of production for one firm, *given free trade*.

However, with the Brazilian tariff policy, which is not likely to be changed owing to Brazil's severe balance of payments problem, and an Argentinian tariff of 35 per cent on CNC lathes, the prices of imported CNC lathes are higher than the international price. This means that, for domestic producers, the minimum efficient scale of production will be substantially decreased. Hence, in a world of tariffs, far fewer than 300 units per year would be necessary. A strategy aiming at mainly penetrating the local markets would therefore put less pressure on the firm to increase the volume of output drastically.

In this scenario, it would also be advantageous for the firm to buy a licence in order to catch up in product design. Of course, buying a licence would also relieve the firm from the need to take risks in design development. A further reason in favour of licensing is that it may reduce the price of the components. In one case, a firm that has a licence from a large Italian firm receives a 20 per cent discount on all the electrical and electronic components that it is forced to buy from the Italian firm. This firm, in turn, has a 30 per cent discount from the producer of the components and makes a profit on the 10 per cent differential. For a firm beginning to produce CNC lathes, an agreement of this kind is most advantageous because an important source of scale advantages, and thus barrier to entry, lies precisely in the discount on components. Hence, the need for large-scale production is considerably reduced with such an agreement.

The correct and perhaps the only possible strategy for the firm was therefore to pursue a short-term strategy of licensing with its marketing efforts directed towards the regional markets. However, we should keep in mind that we are not dealing with a free trade world and that the success of a strategy aimed at exploiting only the regional markets is profoundly dependent on the existence of tariffs or other trade barriers.

Given that licensing was thought to be a part of the firm's strategy, at least in the short term, the question remained to be answered, licence from whom? Two conditions had to be fulfilled by the licensor. First, it had to be a large firm having a production of CNC lathes that permitted it to pay very good prices for the control units. This could, in turn, mean that the licensee could receive some of that discount. Secondly, the firm had to produce smaller and cheaper CNC lathes of a modern design for mass markets, i.e. a firm pursuing the overall cost leadership strategy. Producers from only two countries fulfil these two criteria, namely one Italian firm and several Japanese firms (one German firm's strategy can be only partly classified as the overall cost leadership strategy). As the largest Brazilian machine tool firm has entered into a licensing agreement with the Italian firm in question, a licensor had to be sought in Japan.

Thus, the characteristics of the firm in terms of resources, the lack of a sound reason for a developed country firm to subcontract work to Argentina and the policy of the Argentinian government meant that the firm in 1981 lacked the financial resources to adjust to the requirements set by the nature of competition in the international CNC lathe industry. The first step for the firm to solve its strategy problem was therefore to ask for and receive 35 per cent tariff protection on CNC lathes. The second step was to enter into discussions with a Japanese second-tier firm for a licence. Although no export restrictions were included in the contract, the licensed model was a five-year old one that the Japanese firm had already ceased to export to the US market (Center on Transnational Economy, 1984). The economic value of the lathe in the international market is therefore expected to be very low and one can conclude that strategy number two was chosen. It is of course of interest to note that the government granted tariff protection, previously there had been no tariff on CNC lathes, just at a time when the evolution of the international industry meant that the firm needed to expand production and strengthen its design team. The government policy was therefore directly counterproductive to the goal of creating an internationally competitive industry. Of course, in the dismal economic and political conditions of the time, it could well be argued that granting protection to the firm gave it the necessary basis for survival. It could also be argued that protection made it possible for the firm to stay near the technological frontier instead of falling further behind. Indeed, it is almost a miracle that the firm still exists after the severe crisis in Argentina. In the long run, however, a policy relying on tariffs will only help to create a permanent infant industry.

6.6 Summary and conclusions

The Argentinian engineering sector grew fairly rapidly as a consequence of an inward-looking economic policy between 1960 and 1974. Although the total volume of production is rather small, the structure of the industry suggests that it is fairly advanced. The firms constituting this industry became the market for the Argentinian machine tool industry. The machine tool industry, too, oriented its efforts towards the local market and exported only occasionally, and then only to other Latin American markets. The main government policies responsible for this orientation were tariff protection in conjunction with a wildly fluctuating exchange rate. The latter made an export orientation all but impossible, whilst the former enabled the firms to establish themselves and capture, until recently, around 50 per cent of the local market – mainly for the simpler types of machine tools.

The leading lathe producer followed the same pattern as the rest of the industry. However, the pursuit of a strategy of concentration on the local market, with its small size and wide fluctuations, created a firm that, in terms of resources, is one of the weakest among the CNC lathe producers in the Third World. More specifically, the firm is small, it is weak in terms of design skills, and it has no marketing network abroad. The discrepancy between the actual resources of the firm and the resources required to compete successfully internationally is very large. The firm operated, furthermore, in a local environment characterized by (a) a low demand for CNC lathes and (b) hostile government policies. Finally, the reasons for expecting foreign firms to collaborate on a subcontracting basis do not exist in this industry.

The firm chose in this situation to opt for a strategy of concentration on the regional market on the basis of a licence and tariff protection. Tariff protection was granted by the government and the firm was therefore permitted to continue with low-volume production and a weak design team. It has however survived, which is almost a miracle in the Argentinian context.

Notes

1 $[(165 - 37)/(100 - 37)] - 1$.
2 $[(165 - 74)/(100 - 37)] - 1$.
3 Unfortunately, data on the *value* of investment in CNC lathes are not available, nor

are data on the total value of investment in lathes. However, it is normal that around 20 per cent of the total investment in machine tools is made in lathes. Hence, in 1979, around 20 million USD may have been invested in lathes in Argentina. A reasonable average cost per unit of CNC lathes sold in Argentina is USD 125,000, which implies that CNC lathes to a value of approximately 5 million USD were bought in Argentina in 1979. Thus around 25 per cent of the total value of investment in lathes may have been made in CNC lathes. Hence the diffusion took place on a fairly significant scale.

4 This is an elaboration of unpublished data received from INDEC.

7
The Case of Taiwan

7.1 Growth and structure of the engineering industry

In 1983, the national income of Taiwan amounted to 46 billion USD and per capita income was 2,500 USD. The growth rate in GDP has been substantial in the last two decades: from 1960 to 1980, the annual growth rate (in constant 1976 prices) was as high as 9.6 per cent (CEPD, 1982, p. 23).

The Taiwanese industrial sector also grew at an impressive rate in the 1960s and 1970s, with an annual growth of 9.1 per cent being recorded between 1960 and 1983 (*Statistical Yearbook*, 1984). The growth rate in the 1980s was however reduced somewhat: in the period 1981–83 GDP increased 5.7 per cent per annum (*Statistical Yearbook*, 1984).

The share of manufacturing in GDP was 22 per cent in 1960 and grew to 40 per cent in 1983 (*Statistical Yearbook*, 1984). In contrast to the Argentinian case, however, the engineering sector in Taiwan accounts for a smaller share of the value of production in manufacturing industry. As can be seen in Table 7.1, despite fast growth in the engineering industry, its share of manufacturing is only 26 per cent. Furthermore, the engineering sector in Taiwan is very different from those in the developed countries as well as from that of Argentina. Only about 12 per cent of the value of output has its origin in the Non-Electrical Machinery sector (the true machine building sector), whilst the share of Electrical Machinery, mainly electronics, is very high indeed (see Table 7.2). Hence, Taiwan has a much *lighter* engineering sector than both the developed countries and Argentina.

The high growth rates of industrial production have been generated within the framework of a fairly open outward-oriented industrialization process. Since 1960, when the first moves towards greater openness as a trading nation were taken, exports as a share of GDP rose from 13 per cent in 1961 to 55 per cent in 1983 (see Table 7.3). Exports were also stimulated by a number of measures, although they were not of any considerable magnitude (Liang and Liang, 1981; Patrick, 1981).

Table 7.1 *Production value in the manufacturing industry and the share of the engineering sector in Taiwan, 1976, 1980 and 1984*

Year	Manufacturing industry (current NT b.)	Engineering sector[a] (current NT b.)	Share of engineering sector in manufacturing industry (%)
1976	931	200	21.5
1980	1,994	493	24.7
1984	2,824	739	26.2

Source: CEPD (1985), table 14.
[a] ISIC 381–385.

Table 7.2 *The structure of the value of output in the Taiwanese engineering industry, 1976 and 1984 (%)*

Year	Metal Products	Non-electrical Machinery	Electrical Machinery	Transport Equipment	Scientific and Professional Instruments
1976	22.0	17.0	41.0	16.5	3.5
1984	15.8	11.9	50.7	18.5	3.0

Sources: CEPD (1983), CEPD (1985).

Table 7.3 *The share of exports in GDP in Taiwan, selected years 1961–1983*

Year	GDP (US$ m.)	Exports (US$ m.)	Share of exports in GDP (%)
1961	1,750	225	13
1965	2,800	525	19
1970	5,650	1,675	30
1975	15,368	6,052	39
1980	40,083	21,527	54
1983	49,626	27,263	55

Source: *Statistical Yearbook of the Republic of China* (1984).

The relative openness of the economy also applied to the process of capital accumulation. The share of imports in fixed capital formation in machinery and equipment grew from about 60 per cent in 1960 to 65 per cent in 1965 and, from then until 1980, fluctuated between 66 per cent and 91 per cent (see Table 7.4). The importance of imports of capital equipment is also reflected in Table 7.5, which shows the self-sufficiency ratio in the engineering sector: whilst the ratio was 55 per cent in 1976, it dropped to 52 per cent in 1984.

Hence, the Taiwanese experience is very different from that of Argentina where import substitution took place to such an extent that the local content of investment was 84 per cent in 1973.

Table 7.4 *Investment in machinery, equipment and transport equipment and the import content of investment in Taiwan, 1970–1980*

Year	Fixed capital formation in machinery, equipment and transport equipment (US$ m.)	Imports of engineering products (US$ m.)	Import content of total investment (%)
1970	753	565	75
1971	937	652	70
1972	1,168	898	77
1973	1,672	1,263	76
1974	2,609	2,229	85
1975	2,839	1,868	66
1976	2,871	2,259	79
1977	2,765	2,197	80
1978	3,446	3,123	91
1979	4,755	4,099	86
1980	6,652	5,197	78

Source: CEPD (1982), tables 3–9g, 9–6 and 10–7b.

Table 7.5 *Production and trade in the engineering industry in Taiwan, 1976 and 1984*

Year	Production (NT b.)	Exports (NT b.)	Imports (NT b.)	Apparent consumption[a] (NT b.)	Exports/ production (%)	Imports/ apparent consumption (%)	Self-sufficiency ratio[b] (%)
1976	200	78	98	220	39	45	55
1984	739	447	268	560	60	48	52

Source: CEPD (1983), CEPD (1985).
[a] Production minus exports plus imports.
[b] Production minus exports/apparent consumption.

Export performance is also very different in Taiwan from in Argentina. The export ratio for the Taiwanese engineering industry grew from 39 per cent in 1976 to 60 per cent in 1984. Hence, Taiwan has the same trade pattern as other smaller nations, such as Sweden and Korea, with both high import shares and high export shares. The main exception in Taiwan is the transport industry, where high tariff barriers have been erected. The import share of metal products is also very low, presumably because the simple nature of most products in that sector means that Taiwanese producers can compete well with imports. The trade pattern of different subsectors in Taiwan is shown in Table 7.6.

7.2 The Taiwanese machine tool industry

The Taiwanese machine tool industry has its origins in the period after the Second World War. The early development of the industry was, however, retarded by low local demand as well as by the low technical level of the customers. By the early 1960s, the industry was still a depressed one (Amsden, 1977). By the end of the 1960s, though, a number of larger firms had emerged as demand began to increase. The increasing demand was a function of both growing capital accumulation in Taiwan (Amsden, 1977; and section 7.1) and growing demand for low-quality machine tools in the region of Southeast Asia. (See Amsden, 1977, for an analysis of the competitive strength of the Taiwanese in this region.)

By 1968 the Taiwanese machine tool industry had achieved an export ratio of over 50 per cent (Chou, 1982). Exports at this time were almost exclusively directed towards the regional market in Asia. For example, in 1968, Thailand received 36 per cent, Vietnam 21 per cent and the Philippines over 10 per cent of Taiwan's exports (Chou, 1982). Intensive domestic competition, a small local market and growing regional demand as a consequence of the Vietnam War (Amsden, 1977; Tsai, 1983) could account for the high exports to Asia. A further factor to take into account was the existence of Chinese business communities in the various parts of Asia (Tsai, 1983).

Data on production and trade in machine tools in the period 1969–1983 are shown in Table 7.7. Production rose from 9.2 million USD in 1969 to 211 million USD in 1983. In 1977, Taiwan became a net exporter of machine tools, being the only developing country to have achieved this status. The industry continued its early export orientation and in 1976 the export ratio surpassed 60 per cent and in 1978 it reached a peak of 75 per cent.

Table 7.6 Production and trade in engineering products in Taiwan at sector level, 1976 and 1984

Year	Metal products						Non-electrical Machinery						Electrical Machinery						Transport Equipment					
	P (NT b.)	E (NT b.)	I (NT b.)	AC (NT b.)	E/P %	I/AC %	P (NT b.)	E (NT b.)	I (NT b.)	AC (NT b.)	E/P (%)	I/AC (%)	P (NT b.)	E (NT b.)	I (NT b.)	AC (NT b.)	E/P (%)	I/AC (%)	P (NT b.)	E (NT b.)	I (NT b.)	AC (NT b.)	E/P (%)	I/AC (%)
1976	44	9	3	38	20	8	34	13	42	63	38	67	82	49	31	64	60	48	33	8	20	45	24	44
1984	117	69	6	54	59	11	88	41	86	133	47	65	375	260	125	240	70	52	137	48	25	114	35	22

Source: CEPD (1985).
P = Production
E = Exports
I = Imports
AC = Apparent consumption (production minus exports plus imports)
E/P = Exports/production
I/AC = Imports/apparent consumption

Table 7.7 *Production and trade in machine tools in Taiwan, 1969–1983*

Year	Production (US$ m.)	Exports (US$ m.)	Imports (US$ m.)	Apparent consumption[a] (US$ m.)	Exports/ production (%)	Imports/ apparent consumption (%)
1969	9.2	4.7	9.3	13.8	51	67
1970	10.6	5.6	8.7	13.7	53	64
1971	12.8	6.3	8.3	14.8	49	56
1972	14.9	7.7	9.8	17.0	52	58
1973	22.0	9.3	13.8	26.5	42	52
1974	33.2	16.7	28.6	45.1	50	64
1975	33.0	19.4	26.3	39.9	59	66
1976	44.4	28.4	34.4	50.4	64	68
1977	67.8	47.4	35.0	55.4	70	63
1978	117.8	87.9	54.5	84.4	75	65
1979	189.1	138.2	87.9	138.8	73	63
1980	219.0	159.3	117.7	177.4	73	66
1981	242.3	177.5	93.8	158.6	73	59
1982	n.a.	n.a.	n.a.	n.a.	n.a.	n.a.
1983	211.0	135.4	113.0	188.6	64	60

Sources: 1969–81: data supplied by the Taiwan Machine Tools Producers' Organization; 1983: NMTBA (1985/86).

[a] Production minus exports plus imports.

The developing countries of Asia constituted the main market for Taiwanese machine tools until the mid-1970s. In 1973, these countries accounted for over 69 per cent of the exports of machine tools and 63 per cent of the exports of lathes (Amsden, 1977). By 1978, however, the lion's share of the exports went to North America and to Europe (*Metalworking, Engineering and Marketing*, 1980). The distribution by country of export in 1981 is shown in Table 7.8. Over 77 per cent of the exports went to developed countries that year. Hence, the astonishing growth in the output of machine tools in Taiwan has mostly been based on an export drive to the developed countries, and this process mainly took place after 1975.

Of considerable importance to the Taiwanese firms' ability to switch from serving the Asian markets to the US market was the establishment of a marketing relationship with US distributors. In the mid-1970s, a group of US distributors went to Taiwan to look for producers of cheap machine tools. Some of them opted for establishing subcontracting arrangements with local machine tool builders, while others supplied licences and sold the final product in the USA. The distributors also supplied a great deal of technical information concerning the design of the machine tools and induced the firms to invest in high-quality machine tools, such as surface grinders, which improve the quality and reliability of the final product.

Table 7.8 *The destination of exports of machine tools and lathes from Taiwan, 1981*

Destination	Machine tools		Lathes	
	Value (US$ m.)	%	Value (US$ m.)	%
Developed countries[a] of which:	137.0	77.2	50.8	75.8
North America	93.3	52.6	32.9	49.1
Developing countries	40.5	22.8	16.2	24.2
Total	177.5	100.0	67.0	

Source: Data supplied by the Taiwanese Machine Tool Producers' Organization.
[a] North America, Europe, Japan and Australasia.

It is of interest to note the timing here. At the same time as the developed country firms began to emphasize CNC lathes, the distributors began to seek low-cost producers in Taiwan. These distributors serve, of course, a critical function because they provide the link between the Taiwanese firms, which are often very small, and the final customer in the USA.

Whilst going for an export-led growth in machine tool production, Taiwan has been dependent on imports for the needs of domestic industry for machine tools to a very large extent. The share of imported machine tools in apparent consumption has been above 60 per cent since 1974, with the exception of 1981 when it was just below 60 per cent. The imports have largely consisted of more expensive machine tools, which can be exemplifed in the case of lathes. Unfortunately, the statistics available do not state the number of lathes imported before 1976. However, for that year imports accounted for only 663 units whilst exports accounted for over 13,000 units. The unit value of exported and imported lathes is shown in Table 7.9, which indicates the difference in the types of machine tools exported and imported in Taiwan. In 1977, for example, the average price per unit exported was only 10.7 per cent of the average price of imported units.

Taiwan has clearly concentrated production on low-priced machine tools. The average prices per unit imported into the USA in 1980 of *horizontal engine and toolroom lathes, non-NC,* are shown in Table 7.10, which illustrates the low price of Taiwanese machine tools compared with some major producers. Taiwan has had considerable success in the US market where it was already the fourth largest exporter in 1980. In 1983, Taiwan became the third largest exporter to the USA, a position that it held in 1984 (NMTBA, 1985/86, p. 128).

Table 7.9 *Price per unit of lathes traded in Taiwan, 1969–1984 (US$)*

Year	Export price	Import price	Export price as a % of import price
1969	1,508		
1970	1,470		
1971	1,399		
1972	1,336	n.a.	
1973	1,455		
1974	2,081		
1975	1,759		
1976	1,371	12,518	10.9
1977	1,744	16,283	10.7
1978	2,029	10,860	18.7
1979	2,739	15,193	18.0
1980	2,935	18,266	16.1
1981	3,328	15,525	21.4
1982	3,234	20,617	15.7
1983	2,601	20,626	12.6
1984	2,671	28,432	9.4

Sources: 1969–81: data received from Taiwan Machine Tool Producers' Organization; 1982–84: data received from MIRL.

Table 7.10 *Price per unit of horizontal engine and toolroom lathes, non-NC, in US imports, 1980*

Origin	Price per unit (US$)	No. of lathes	Value (US$ '000)
All imports	9,243	8,023	74,164
Taiwan	5,386	3,118	16,795
Japan	36,054	437	15,756
Italy	18,394	152	2,796
Poland	10,792	467	5,040
UK	7,828	995	7,789

Source: Elaboration of data supplied by NMTBA.

7.3 Government policy

A major instrument of government policy in Taiwan has been the tariff system. The structure of tariffs is, however, complex and the rates vary from above 100 per cent to zero, although the average legislated rate of *dutiable imports* is not higher than 15–17 per cent (Westphal, 1978, p. 19).

Concerning the general industrial policy, which would also apply to machine tools, Westphal (1978, p. 37) notes, that since 1969 'the effective subsidy to manufactured exports must have declined to now negative or only marginally positive, as the credit subsidies which remain seem unlikely to compensate for taxes and tariffs paid indirectly through the purchase of non tradeable inputs and for the increase in costs due to import controls'. Although Westphal (1978, p. 40) also suggests that 'relative incentives have changed to increase the encouragement to import substituting activity at the expense of exporting activity', the evidence points to only a relatively weak bias in favour of sales to the domestic market.

For the machine tool industry, the picture appears to correspond broadly to the average in the manufacturing industry. According to several Taiwanese sources (Chou, 1983; Tsai, 1983), no special government policy vis-à-vis the machine tool industry existed in the past. Amsden (1977), too, notes that there had been hardly any government aid to the machine tool industry. Table 7.11 shows the nominal rates of protection for lathes for the period 1948–81. The rates are low and correspond roughly to the average in Taiwan (see Westphal, 1978, p. 17). The effective tariff rates would be roughly the same as the nominal ones on account of the very low import content of conventional machine tools and the international prices of import raw materials in Taiwan. Indeed, firms claim that products such as castings are cheaper in Taiwan than in the rest of the world. Only one product was, according to one firm, more expensive in Taiwan in the mid-1970s, namely alloy steel. The effect on, say, the price of an engine lathe was, however, negligible (interview).

Table 7.11 *Nominal tariffs for lathes in Taiwan, selected years, 1948–1981 (%)*

Year	1948	1955	1960	1965	1971	1976	1981
%	7.5	7.5	9.0	9.0	13.0	13.0	10.0

Sources: 1948–76: Scott (1979), table 5.2. 1981: Industrial Development and Investment Center (IDIC) (1982b).

The overall slight general bias in favour of the domestic market does not appear to have affected firms' decisions as regards sales to export and domestic markets. As was shown in section 7.1, exports began at an early phase of the development of the industry. The orientation to the export market, in spite of the overall policy bias, is not surprising as other considerations need to be taken into account when analysing a firm's decision where to direct its sales efforts. I noted above that the domestic market was saturated at an early stage and that there was intense competition. Barriers to entry were very low. This, of course, led to an interest in the export market.

Another major instrument of government intervention is import licences. These also exist for machine tools: a machine tool is allowed to be imported if it is not produced in Taiwan. A neutral body, i.e. neutral in relation to the machine tool industry, decides which machine tools can be imported. As a very substantial import of machine tools exists (see Table 7.7), this body would seem to have had little effect on the level of protection.

As noted in Chapters 4 and 6, exporting can be a risky enterprise and a key factor to consider is the uncertainty created by the exchange rate policy of the government. I noted that the real exchange rate in Argentina varied tremendously over the years, and indeed even between different quarters. Taiwan, in contrast, shows a stability in the real exchange rate as can be seen in Table 7.12, where the Taiwanese rates are contrasted with the Argentinian.

Table 7.12 *Fluctuations in real exchange rates in Taiwan and Argentina, 1968–1979*

Year	% change on previous year	
	Taiwan	Argentina
1969	3.6	− 2.8
1970	2.0	− 3.9
1971	3.9	−10.7
1972	3.9	31.4
1973	− 5.8	−26.3
1974	−17.3	− 6.3
1975	14.1	32.3
1976	3.5	− 7.0
1977	2.1	14.4
1978	2.2	−22.2
1979	n.a.	−41.9

Sources: Taiwan: Liang and Liang (1981), table 4; Argentina: Table 6.8.

Hence, the government policies in Taiwan have been roughly neutral as regards both export and domestic markets: it has created stability as regards the relative prices of domestic and foreign goods by maintaining a fairly stable real exchange rate, and the policy of outward-oriented growth has been associated with an accumulation of capital in Taiwan, thereby creating a steadily growing domestic market for machine tools.

7.4 The diffusion of CNC lathes in Taiwan

In this section, I shall briefly present the actual rate of diffusion of CNC lathes in Taiwan between 1977 and 1984 and then discuss future demand for CNC lathes.

The production, trade and apparent consumption of CNC lathes in Taiwan for the period 1977–84 is shown in Table 7.12. The share of CNC lathes in total investment in lathes is also indicated. As in Argentina, sales of CNC lathes were very small in 1977 and 1978 but they took off in 1979 and were around 100 units in 1981–83. In 1984, they increased to 250. The share of CNC lathes in total investment in lathes (in value terms) also grew from a mere 6 per cent in 1978 to around 20 per cent in 1980 and 1981 only to increase to an astonishing 53 per cent in 1984. Although Taiwan experienced a rate of diffusion of CNC lathes in 1984 equivalent to that in many developed countries in 1980, the market was still very limited measured in terms of units sold.

Table 7.13 *Production and trade in CNC lathes in Taiwan, 1977–1984 (in units and value)*

Year	Production		Exports		Imports		Apparent consumption[a]		Share of CNC lathes in total investment in lathes	
	Units	Value (US$ '000)	Units	Value (US$ '000)	Units	Value (US$ '000)	Units	Value (US$ '000)	Units %	Value %
1977	14	459	10	303	15	1,126	19	1,282	0.7	7.9
1978	40	1,512	29	985	19	1,035	30	1,562	1.0	6.0
1979	78	3,397	54	2,203	59	3,326	83	4,520	2.4	12.2
1980	106	4,718	64	2,929	95	4,326	137	6,115	4.1	19.9
1981	174	6,386	106	4,461	55	3,546	123	5,471	2.8	19.4
1982	131	4,960	100	3,939	49	4,289	80	5,310	2.5	22.0
1983	144	5,684	82	3,278	79	5,063	141	7,469	4.2	28.5
1984	347	13,754	206	7,438	109	6,831	250	13,147	6.3	52.8

Source: Data received from ITRI and MIRL.
[a] Production minus exports plus imports.

What of the future though? Below I shall present an estimate of the demand for CNC lathes in 1982–86. The method involves making assumptions about three variables: the total investment in lathes between 1984 and 1988; the share of CNC lathes in this investment; and the price of each CNC lathe. First, I assume an annual 10 per cent growth in real terms in investment in lathes in these four years. I also assume that the share of CNC lathes in total investment in lathes will grow linearly to reach 60 per cent in 1988. Finally, I assume that the price of CNC lathes will be the average of the prices of CNC lathes in Taiwan in 1982–84 (55,000 USD). Using these assumptions, we would see a growth in the demand for CNC lathes in Taiwan from 250 units in 1984 to 398 units in 1988.

Thus, even assuming both an increase in the share of CNC lathes in total investment in lathes *and* a 10 per cent annual growth in the investment in all lathes, the size of the home market will be less than is required for one firm to reach the break-even point. Hence, even a rapid growth of user industries and a rapidly rising share of CNC lathes in the total investment in lathes would not create a large domestic market for CNC lathes in Taiwan.

7.5 The firms producing CNC lathes

About thirty firms produce lathes in Taiwan. The total production of lathes grew dramatically from 3,228 units in 1969 to 23,935 units in 1981 only to decrease to 20,726 in 1984 (see Table 7.14). As in the case of all machine tools, the mid-1970s was the time of the greatest expansion.

Exports of lathes have been high – fluctuating around 60 per cent of the output (in terms of units) until the mid-1970s when the export share rose to over 80 per cent of production. In terms of units, the import share of investment is low – less than 20 per cent in the 1980s.

In terms of value, the picture is a bit different. The export share is still high, although slightly lower than if we measure in terms of units (see Table 7.15). The import share of investment is, however, substantially higher than if we measure in terms of units, indicating a specialization in the production of low-value units in Taiwan. (See Table 7.9, too.) The import share of investment fluctuated around 50 per cent until 1978 when it fell, and it now fluctuates around 35 per cent.

The firms producing lathes are generally small, reflecting the low barriers to entry into the production of engine lathes. The average size firm is, however, much larger than its Argentinian counterpart. Whereas the twenty Argentinian firms share a production of around 2,000 units, thirty Taiwanese firms produce over 20,000 units. Three firms account, however, for about 30 per cent of the value of production. Two of these firms constitute the technically leading firms.

Out of the thirty lathe producers, a minimum of five firms have begun to produce CNC lathes. These five firms include the three largest firms and two firms producing a smaller number of lathes. Another firm, which does not produce conventional lathes at all, has also begun to produce CNC lathes. Two of the three small firms belong

Table 7.14 *Production and trade in lathes in Taiwan, 1969–1984 (units)*

Year	Production	Exports	Imports	Apparent consumption[a]	Exports/ production	Imports/ apparent consumption
	Units	Units	Units	Units	%	%
1969	3,228	2,025	n.a.	n.a.	63	n.a.
1970	3,554	2,433	n.a.	n.a.	69	n.a.
1971	4,569	2,664	n.a.	n.a.	58	n.a.
1972	5,365	3,591	n.a.	n.a.	67	n.a.
1973	6,490	3,434	n.a.	n.a.	53	n.a.
1974	7,707	4,804	n.a.	n.a.	62	n.a.
1975	7,933	4,775	n.a.	n.a.	60	n.a.
1976	13,068	8,895	663	4,836	68	14
1977	14,284	11,925	522	2,881	84	18
1978	17,649	15,722	930	2,857	89	33
1979	21,510	18,799	724	3,435	87	21
1980	22,232	19,521	646	3,357	88	19
1981	23,935	20,132	599	4,402	84	14
1982	16,411	13,623	397	3,185	83	12
1983	15,449	12,561	461	3,349	81	14
1984	20,726	17,120	347	3,953	83	9

Source: Data supplied by the Taiwan Machine Tool Producers' Organization and MIRL.
[a] Production minus exports plus imports.

to larger engineering firms whilst one is a small independent machine tool producer.

I shall describe the history of each of these firms briefly and then discuss their present position within the industry producing CNC lathes.

(i) Firm A (refer to Figure 5.1) The firm was established in 1943 as a textile machinery producer. In the second half of the 1960s the firm decided to move into the production of machine tools and in 1967, it developed Taiwan's first precision high-speed lathe, an engine lathe. The growth of sales of these lathes is shown in Figure 7.1. In 1980 the firm produced 2,300 engine lathes. This number can be compared with the total Argentinian production of lathes of about 2,400 units in 1979 (see Table 6.12). Furthermore, the Taiwanese firm produced about 110 NCMTs in 1980 as well as 400 fork-lift trucks. Total sales reached 15.4 million USD in 1979 and 19.5 million USD in 1981 (Annual Reports). In 1984 production reached 150 CNC lathes and 90 machining centres. This level of sales makes it the leading machine tool builder in Taiwan.

Table 7.15 *Production and trade in lathes in Taiwan, 1969–1984 (value)*

Year	Production	Exports	Imports	Apparent consumption	Exports/ production	Imports/ apparent consumption
	(US$ m.)	(US$ m.)	(US$ m.)	(US$ m.)	(%)	(%)
1969	4.7	3.1	1.8	3.4	66	53
1970	5.1	3.6	1.7	3.2	71	53
1971	6.4	3.7	2.0	4.7	58	43
1972	7.6	4.8	2.3	5.1	63	46
1973	10.8	5.0	3.3	9.1	46	36
1974	16.9	10.0	5.4	12.3	59	44
1975	13.7	8.4	4.8	10.1	61	48
1976	18.2	12.2	8.3	14.3	67	58
1977	28.7	20.8	8.5	16.4	73	52
1978	48.0	31.9	10.1	26.2	67	39
1979	77.5	51.5	11.0	37.0	67	30
1980	76.1	57.3	11.8	30.6	75	39
1981	86.0	67.0	9.3	28.3	78	33
1982	59.9	44.0	8.2	24.1	74	34
1983	49.4	32.6	9.5	26.3	66	36
1984	60.8	45.7	9.8	24.9	75	39

Source: As for Table 7.14.

The firm only produced engine lathes until 1974. At first it copied a Japanese model but later it designed its own engine lathe. In 1974, the firm developed its first CNC lathe. The development of sales of CNC lathes is shown in Figure 7.2.

In the mid-1970s, the total number of designers were thirteen. Between 1975 and 1983, this number grew steadily to thirty-five. The growth in designers was accompanied by an intensive design development effort. These designers were helped by the acquisition of a Computer Aided Design unit, which was installed in 1980. In 1979, the first vertical machining centre was developed. 1980 saw the birth of a modern CNC lathe of its own design, which was a very fast response to the design changes originating in Japan. A second machining centre was also put on the market in 1980. In 1981, another CNC lathe of the same basic design as the previous one was made, together with yet another machining centre. By 1985, the firm had three CNC lathe designs and four machining centre designs of its own make. Together with firm B, this firm is the first NIC-based firm to have a complete series of CNC lathes.

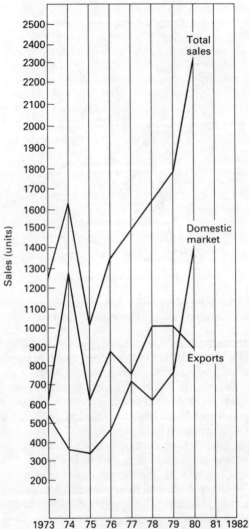

Figure 7.1 *Sales of engine lathes for firm A, 1973–1980*
Source: Booklet distributed by the firm.

As is shown in Figure 7.1, exports exceeded sales on the domestic market for engine lathes, except in 1980; in the case of CNC lathes, exports account for an overwhelming part of the production. The overall export ratio was 60 per cent in 1985 and the bulk of the exports are now accounted for by CNC lathes and machining centres. As was

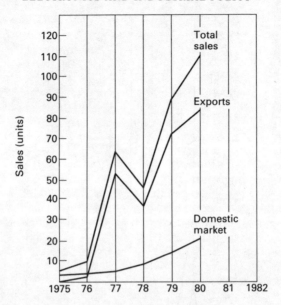

Figure 7.2 *Sales of NC machine tools for firm A, 1975–1980*
Source: Booklet distributed by the firm.

the case with the lathe-producing industry as a whole, this firm took the US market as its home market in the 1970s. In 1973, the director of the firm met with a US distributor which subsequently handled its sales until recently. Sales were directed towrds smaller firms, e.g. tool and die shops, which were highly price sensitive. Hence, the product/market combination of the firm was cheap, mass-produced engine lathes for the smaller users in the USA.

Similarly, the CNC lathes were designed with the purpose of achieving low production costs and are aimed at the jobshop market in the USA and to some extent the same niche in Europe. The low-performance strategy discussed in section 5.1 was clearly chosen as a strategy of entry by this firm. The firm has thus chosen an international market niche characterized by high price sensitivity (in sharp contrast with the leading Argentinian firm).

As was indicated in Figure 5.1, the firm is the leading one among the eight firms surveyed in the three NICs. It is of reasonable size, with 700 employees and sales of slightly more than 19 million USD in 1981. The firm has invested in a sales and service centre in the USA. In 1982 it discontinued its collaboration with the distributor firm that had been its agent for ten years. The reason for this change was that the agent could not handle the more complex machines that the firm was moving into and could not keep up with the growth rate. So the firm invested $3–4 million in a centre employing twenty-five people, sixteen of whom are service engineers. The firm is also beginning to build up an organization in the West German market. Other markets are served by independent sales representatives. Finally, the firm has a reasonably sized design staff and, among the NIC-based firms, is probably the strongest in terms of design skills. The present strategic position is one of a consolidated entry into the low-performance strategy.

(ii) Firm C The firm was founded in 1956 and initially produced a range of simple machine tools, such as belt lathes. For a long time it was a very small firm – in 1971, sales amounted to only 200,000 USD. By 1975, sales had increased to 800,000 USD and the firm developed its first high-speed precision lathe, eight years after the market leader. Exports and production increased rapidly and, by 1980, sales had increased to nearly 10 million USD and in 1981 the sales of high-speed precision lathes in the US market alone reached 3,000 units.

Up until 1981, the firm produced only engine lathes. In 1981, however, the firm developed a CNC lathe of an old concept (see Chapter 5) and sold 23 units, as well as 7 units without the CNC unit attached to it. In 1982, the firm presented a CNC lathe of modern concept that took one and a half years to develop. It is, however, a copy of a well-known Japanese CNC lathe. In 1982, 55 units of this CNC lathe, as well as 5 incomplete units, were produced. In 1982, the firm also developed a machining centre. A second modern CNC lathe and another machining centre were developed in 1983 or 1984. Production in 1984 amounted to 122 CNC lathes and 16 machining centres. Prior to this development effort, the firm had only ten designers. With the beginning of the CNC era for the firm, it expanded the design staff to twenty to twenty-five engineers. The firm devoted 7 per cent of its sales to R&D in 1983, but this figure was expected to decline somewhat in the following year.

The firm has always been export-oriented and has had an export ratio of 70 per cent for many years. The export ratio for the CNC lathes was 50 per cent in 1982. Like the other Taiwanese firms, this firm aims at the smaller customers in the developed country markets. For CNC lathes, the firm has clearly chosen the low-performance strategy.

In Figure 5.1, the firm was ranked as number three among the NIC-based firms that have been surveyed. It is a reasonable-sized firm, which in 1982–83 according to firm sources had sales of 15 million USD. The firm has recently, as mentioned above, also increased its design team significantly, and has invested in three firms in the USA for the sales and servicing of machine tools. These employ thirty people, nine of whom account for service and sales of their own machine tools while the rest sell other Taiwanese-made machine tools.

The firm was making a loss on the CNC lathes it produced in 1982–83. In order to consolidate its entry, the firm needed to both increase volumes and slightly improve its design capabilities while moving away from the copying stage. In other words, it needed to move towards northeast in Figure 5.1. In 1984, the firm seems to have achieved a fairly large volume of output; indeed it was 2.5 times as large as that in 1982. Further expansion is, however, needed. The plans of the firm take this into consideration. It is now constructing a new plant with a larger production capacity. The plant is partly financed by the Bank of Communication within the framework of the Strategic Industry Programme (see section 7.6 below). With the capital available from the Bank, the firm would seem to be able to make the required investments and successfully mature and consolidate its entry into the low-performance strategy, as well as begin to shift over to another strategy.

(iii) **Firm F[1]** This firm was established in 1949 as a manufacturer of bicycle wheel rims. Over the years, the firm diversified into a large number of engineering products such as steel pipes, castings, machinery for a number of products and machine tools. Sales grew from 2.7 million USD in 1971 to 47.3 million USD in 1981. Less than 20 per cent of sales are accounted for by machine tools.

The firm concentrates its production of machine tools on larger items, such as horizontal boring mills and produces no conventional lathes. In 1980, the firm began designing a CNC lathe of a modern design. The designers bought in two Japanese CNC lathes, dismantled them and took the best parts from each machine tool. The firm now has two models of the same basic concept and sold 6 units in 1982, all on the export market. The firm is developing a third model of the same basic concept. The firm has thirty designers, but only a few of these are involved in designing machine tools.

The firm's strategic position is rather different from the other firms in that the director aims to use the CNC lathes to get a reputation of reliability abroad. The CNC lathes are in the low–medium performance bracket and the firm loses money on each unit sold. The director counts on a break-even point of 40 units per month, but he does not think that they will manage that many. They have, however, set up a service centre in the USA (1977). The firm's aim is to sell its large machine tools, where little scale economies can be found, and to sell ancillary equipment for FMS on the foreign market. Hence, the objective of the firm is not primarily to develop the CNC lathe market, even though its financial position and to some extent marketing capabilities would mean it has the possibility of aiming at a more ambitious goal.

(iv) Firm G The firm was established in 1970 as a producer of engine lathes and bench lathes. The firm employed ninety people in 1983, although a year before the number was 150.

Until 1980, the firm did not produce CNC lathes, but then it copied a Japanese CNC lathe of a very old design (see Chapter 5). It is now developing a model of its own design, although still of a very old concept. The firm has five designers. In 1982, it produced 16 CNC lathes.

Ever since the establishment of the firm, it has been export-oriented (currently 90 per cent of sales). Initially, the Southeast Asian market was the main market but in 1977 a distributor from the USA offered to sell the firm's lathes in the USA if it improved the quality in various ways. The USA is now the main market (50 per cent of sales). As in the case of the other firms, this firm aims at the market of small shops, repair shops, etc. The firm is sticking to the very old concept for CNC lathes because the director (probably erroneously) claims that this type of low-performance CNC lathe has a big market in the USA.

The present strategic position of the firm is one of trying to enter the market for CNC lathes by choosing a very particular segment in the US market. Its entry is helped by the existence of its own organization in the USA, which services and sells its machine tools to regional distributors. The firm is, however, making a loss on its CNC lathes and, owing to its weak design department and small size, it would seem difficult for it to consolidate its entry without an external injection of funds. Alternatively, the firm could gradually grow on the basis of its engine lathes and only later move into CNC lathes on a larger scale. In comparison with the other firms, this firm is far behind.

(v) Firm H The firm was established in 1918 as a construction and civil engineering firm. Today, the firm is a large conglomerate with its base in the electrical/electronic area. Its sales amounted to 526 million USD in 1981. The firm also has a number of subsidiaries in other countries, such as the UK and in USA.

In 1974, the firm started machine tool production. It began by licensing a small surface grinder from Japan. In 1976, it started engine lathe production, again with a licence from one of the largest Japanese machine tool builders. In 1982, it produced about 20 units per month and sold back 95 per cent to the licensor. In 1981, the firm embarked on the production of a CNC lathe with a licence from the same firm. The model is, however, of an old concept that was popular in Japan in the second half of the 1970s. Only 5 units were produced in 1982 and all were for internal use. As the firm cannot get a modern design from the Japanese firm, it has started to develop a CNC lathe by copying a modern design.

Before the copying began, the firm had six designers in machine tools out of 250 employees. These designers developed an engine lathe design and a grinding machine design. It was realized that more designers were needed, so a task force was created with up to ten engineers, who work on the CNC lathe design on a part-time basis.

The strategic position of the firm is that it is in the process of trying to enter into the market for CNC lathes. Licensing has proved to be unsuccessful, partly owing to the refusal of the licensor to supply modern designs. The licensor also claims that the licensee is too weak technically to absorb knowledge (interview). The firm is developing a new model but, as the specifications of the design are not available, it is not possible to ascertain the market segment for which it is aiming. The firm's machine tool division is, however, very weak in skills as a consequence of the earlier licensing, and a substantial investment in skills would be required for the firm to be successful in its entry into the low-performance strategy. The conglomerate to which the firm belongs is very large and has both the financial resources and the marketing network required for the machine tool division.

In early 1983, there were also rumours that Fanuc, the Japanese CNC builder, would link up with this firm for the production and development of CNC units in Taiwan. If this took place, a competitive edge could be gained by the machine tool division in CNC lathes on account of access to cheap control units.

(vi) Firm I (this firm is not included in Figure 5.1) This firm was established in 1950 and began machine tool production in 1971. Its sales today are 12 million USD, of which half is in machine tools.

The firm began producing engine lathes and drilling machines under licence from a Japanese firm, but about ten years ago it initiated its own design development. In 1979, it began designing a CNC lathe of an old concept. No sales were made though, and the firm has just developed a four-axis CNC lathe of a modern design (a more complex CNC lathe).

The firm has a total of twenty-five designers, half of whom are in machine tools. Upon starting the design development of the modern CNC lathe, it hired five new designers and sent them to Japan for training. The firm also received a loan of several million USD from the Bank of Communication for the CNC lathe programme. The loan was partly to buy grinding machines so as to ensure high quality.

The firm is a fairly small one with a relatively weak design department, even though it is stronger than several other firms in Taiwan. The firm also has an independent company in USA for sales to distributors where ten people are employed. The availability of the marketing function in the USA improves the firm's position and the

THE CASE OF TAIWAN   163

emphasis on a rather unusual product, the four-axis lathe, ensures that the firm is not competing with all the other firms for the low-performance segment.

(v) **Summary** As could be seen in Table 5.4, the firms surveyed vary a great deal with respect to a number of characterisics, such as number of designers, origin of the designs, as well as number of employees. They also differ in terms of their institutional characteristics – most are independent but one is a division of a very large Taiwanese firm. The firms also vary with respect to their strategic position. The leading firm is in the position of attempting to shift strategy to the overall cost leadership strategy, while several other firms are just entering into the production of CNC lathes. Furthermore one firm's aim in producing CNC lathes is to use it as a marketing tool only. The firms share, however, the characteristic of exporting the majority of their output. The exports are concentrated to the US market, where all of them have some form of marketing subsidiary.

7.6 Government policy in the machine tool field

The Taiwanese industry has gone through several phases in the past twenty years. In the early 1960s, when the first export orientation took place, light industries such as consumer electronics and home appliances were emphasized. The 1970s saw the beginning of backward integration and the development of petrochemical industries. Some items in the machinery industry, such as machine tools and motor vehicles, were also developed (CEPD, 1981).

One way of showing the kind of structural change that has taken place in Taiwanese industry is the ratio between the value of output in the traditional consumer goods industries and the value of output in the engineering industries (see Table 7.16). The ratio declined from 4.6 in 1965 to 1.6 in 1982.

For the 1980s, the stated objective of the Taiwanese government is to 'promote the development of strategic industries and accelerate the structural transformation of industry' (CEPD, 1981, p. 8). In order to accelerate the structural transformation of the industry, industries should be chosen that:

- have a low energy intensity
- have high value added
- have high linkage effects
- have a high skill content.

Table 7.16 *The ratio between consumer goods[a] and engineering goods[b] production in Taiwan, 1965 and 1982*

Year	Ratio
1965	4.6
1982	1.6

Sources: CEPD (1975); CEPD (1983).
[a] Defined as food, beverages, tobacco, textiles, garments, wood and wood products, paper and paper products, printing and leather.
[b] Defined as ISIC 381–385.

The Taiwanese government designed the Strategic Industry Programme in 1982. The programme consists of a number of policies (dealt with in more detail below) applying to around 150 products chosen by the government in collaboration with industry and the Industrial Technological Research Institute (approximately 100 people were involved in the choice of the 150 products). The scope of the programme is:

- forty-nine products in the non-electrical machinery industry, e.g. NCMTs, industrial robots, Flexible Manufacturing Systems (FMS), Computer Aided Design units (CAD) and computer-controlled sewing machines;
- nine products within the automobile components category;
- twelve products classified as electrical machinery;
- sixty-five products within the information and electronic industry, e.g. CNC units (IDIC, 1982).

The programme includes four components: tariff policy, provision of risk capital, export incentives, and industrial R&D. I shall discuss each component below.

7.6.1 Tariff policy

In 1983, the tariff rate for the 150 products was 10 per cent. This also applied to CNC lathes and other lathes. In 1983 it was planned to increase this tariff rate to 20 per cent for a period of three–five years with the aim of giving local suppliers the opportunity to exploit the local market and increase volume of output.

7.6.2 Provision of risk capital

The state-owned Bank of Communication was allocated 10 billion NT dollars (approximately 250 million USD) for this programme. The Bank evaluates applications and may grant money to individual firms that produce or plan to produce a product that is classified as strategic.

The screening procedure is said to be rather tough: the Bank studies the profits from the past three–five years and the capabilities of the management; does a technical evaluation; and studies the financial side of the project. The stated objective is to specify the *social* value of the investment project. In an interview, a high-ranking official in the Bank strongly suggested that *laissez-faire* is not good for developing countries; he was convinced that a small group of educated officials have better information than most firms and that they are able to pick winners and allocate scarce capital better than the individual firm. This philosophy lies behind the whole programme. The programme lends capital to firms at a 2 per cent lower rate of interest than the prevailing one. The lower interest rate is primarily justified by the low levels of investment given the worldwide recession. It is thus a capital subsidy. However, the main function of the programme is not to subsidize capital in the normal sense but to supply risk capital without a high-risk premium. According to the Bank, normal banks are very conservative and have a high risk aversion. Given that a change in the structure of industry necessarily implies moving into new products, often of a more complex nature than those previously produced, high risks may prevail at the level of the firm. The Bank of Communication (or the state) therefore absorbs some of the risks, in that it pays a maximum of 65 per cent of the capital involved in an investment.

7.6.3 The Export–Import Bank

The Export–Import Bank was founded in 1979 with the specific objective of aiding Taiwanese exports of machinery. The Bank is state-owned and has a number of programmes.

The main function of the Bank is to provide medium- and long-term credit. In 1981, 82 million USD (54 per cent of the total disbursement) was disbursed for this type of credit. The main recipient was the shipbuilding industry. 'Machine tools and others' accounted for only 0.3 per cent of the total disbursement. Turnkey plants received 6 per cent and industrial machinery 0.5 per cent (Export–Import Bank of China, Annual Report, 1981).

The Fixed Rate Relending Facility (FRRF) is a specific credit line that makes credit available to selected foreign financial institutions for relending to their clients in order for them to buy non-project-related machinery where each transaction amounts to less than 1 million USD. The bank charges an interest rate of 8.5 per cent p.a. and the relending institution is allowed to add another 1.5 per cent p.a. The outstanding funds within this programme grew from 330,000 USD at the end of 1980 to 3 million USD at the end of 1982. Nearly 50 per cent went to the USA, Canada received 5.7 per cent and Australia 2.7 per cent – all these figures being for machine tools only.

The Bank also has a scheme whereby it provides pre-shipment financing to exporters upon receiving the Letter of Credit. The interest rate is the same as the one for general export loans, or 8.5 per cent p.a., which is 1–2 per cent lower than the general interest rate.

An import credit facility exists whereby the Bank assists firms that need extra manufacturing time, for example in shipbuilding, or firms that need financial help with the purchase of raw materials. This scheme accounted for 43.8 per cent of the credit disbursement in 1981 (Export–Import Bank of China, 1981).

Finally, the Bank runs an export insurance scheme whereby exporters are protected against political and commercial risks overseas. The Bank also has a guarantee programme aimed mainly at the shipbuilding industry.

7.6.4 R&D on automation

The Mechanical Industry Research Laboratories (MIRL) is the institution responsible for the government Programme on Production Automation. The MIRL is, however, only partly funded by the government and is partly a commercial institution. It was founded in 1974 with the specific purpose of helping the local machine tool industry to upgrade its quality. At this time, the local machine tool builders produced for the local and regional markets, which demanded low-quality machine tools (see Amsden, 1977). The first programme was to introduce numerically controlled machine tools to local machine tool builders. This was done by buying three NCMTs and displaying them at local exhibitions. MIRL also sent four engineers abroad to study NCMTs – how to use and produce them. The second step was to begin the production of machine tools under licence. The purpose was twofold: first to acquire a bread and butter machine; second, to learn how to produce advanced machine tools. In connection with this production it trained forty to fifty engineers and gave local industrialists advice. By 1980, MIRL had achieved the basic skills and decided to go for higher technology.

MIRL now has 120 mechanical engineers and 60 electronic engineers whom it uses in a variety of programmes. For example:

In 1980, all precision gears were imported from Japan, so MIRL sent a proposal to the government to study how to make precision gears. The government paid for the very expensive machinery, whilst MIRL paid for the engineers. It sent six people abroad to learn how to make precision gears and by 1983 it had the capacity to supply 20 per cent of the domestic market. It also sent engineers to two local firms to assist them in the production of gears. It claims that the local supply covers 50 per cent of the market and that Japanese gears are now 30 per cent cheaper than previously.

About ten CAD systems were installed in Taiwan by 1983. As this is an important tool for raising the productivity of designers, MIRL has acquired two CAD units for testing and for teaching. It is also allowing local firms to use the CAD units on a temporary basis.

An important part of the activity of MIRL is to design advanced machine tools for local firms. By 1983 it had designed two CNC lathes and one robot for a local machine tool firm, where the firm paid for the materials and MIRL for the engineers. It has also designed a machining centre and a CNC grinder. Finally, it has designed and produced an FMS consisting of a number of machining centres equipped with automatic pallet-changers and a robot-trailer. This is a very advanced piece of technology, which does not exist in many parts of the world. The system was designed for a local producer of car wheel axles whose production volume did not justify the acquisition of specially designed machine tools. MIRL also designs special machine tools at a price that it claims to be many times lower than the international price.

MIRL also produces some high-precision components, such as headstocks for CNC lathes, on a subcontracting basis for local firms.

Finally, it has a large programme for developing a local computer numerical control unit. It has a long-term strategy that aims at commercializing a Taiwanese CNC unit on a large scale. In 1978 it began developing a point-to-point CNC (a simple one). By 1983 it had sixty electronic engineers, although many are inexperienced. The development work is very difficult as MIRL claims that there is no information available, in the sense that it has nothing to copy. It plans, however, to supply the first CNC units in 1986 for use on the local market. It intends to give the unit to a local computer firm that it will continue to subsidize with development work to ensure early commercialization. The main costs involved are development work so the scale economies are substantial, but if the continuous upgrading is done by MIRL the firm can start producing at a rate of only 10 units per month. Only at a later stage are there plans for a reduction in the subsidy.

7.6.5 China External Trade Development Council

As a final element in explicit government policies, I need to mention the China External Trade Development Council (CETDC). In 1970, the government provided the initial money for the creation of CETDC, which employs around 300 people, seven of whom are in the machinery division. The CETDC is now financed by the exporters themselves through an export tax. The main function of CETDC is to act as a matchmaker between local firms and foreign firms. CETDC does market research abroad, it assists local firms to participate in foreign fairs and, together with dealers of Taiwanese-made machinery abroad, it helps to establish *permanent* machinery exhibitions. Twenty such exhibitions currently exist in the machinery industry, and many of these are machine tool exhibitions.

As there are large indivisibilities in the gathering of information in external markets, an organization of this kind can be very helpful.

7.7 Evaluating the explicit governmental policy

The explicit governmental policy covers four areas: tariffs, risk capital provision, export financing and centralized R&D. I also briefly mentioned a government-initiated marketing institution, CETDC. In the following sections I shall assess these government policies in the light of the earlier discussions on the nature of competition in the international CNC lathe industry and the present position of the six Taiwanese firms within this industry.

The role of government policy within the process of capability generation will be discussed further in Chapter 9. I shall just state here that, under some conditions, the government may be justified in intervening in this process. The essence of government intervention can be seen, during a shorter time period, as helping firms that are attempting to change strategy, or product/market mix, to bridge the gap between their present resources/capabilities and the resources/capabilities required to pursue the new strategy.

7.7.1 Tariffs

The plan was to raise the tariffs for all products, including CNC lathes, to 20 per cent. I shall argue that such an increase would be both inefficient and of marginal importance to the CNC lathe producing firms.

(i) Inefficient The tariff policy proposed by the Strategic Industry Programme is a *general* instrument and is therefore blunt. If we take the Taiwanese firms discussed in section 7.4, a tariff rate of 20 per cent would benefit any of these firms as long as they sold on the local market. To the extent that the price level would increase, thus representing a tax on the users of machine tools, resources would be channelled into, for example, firm A in Figure 5.1. This firm has already consolidated its entry into the low-performance strategy and operates within a context of fairly low risks. It does not therefore require any external funds, at least not until it wants to change strategy (see below). Not only would this firm receive taxed money, but so would firms that really do not seem to be capable of consolidating their entry within a reasonable time period, e.g. firm G. Indeed, given that the break-even point may on average be at least 400 units per annum (using the very low estimate provided by the Boston Consulting Group, 1985), it is not reasonable to assume that all the firms that are attempting an entry should be able to make it. If all these firms were to reach break-even point in, say, five years, it would mean an annual production of at least 2,400 units by 1990, which would mean an average annual growth of Taiwanese production of CNC lathes in the order of 50 per cent. This is not judged to be feasible, so this tariff is bound to channel resources into some firms that will not be successful anyway. This issue will be discussed more in detail in Chapter 9.

The government needs to take into consideration questions of economic efficiency as well as questions of growth in technological and industrial competence. In other words, with scarce resources, a government should not allocate funds either to firms that do not require them or to firms that stand very little chance of succeeding in the strategic change, i.e. firms where the gap referred to above is too large to be bridged within a limited period of time.

(ii) **Of marginal use**   The official argument in favour of tariffs is that, by partly excluding foreign CNC lathes from the domestic market, the firms would be able to reach a higher volume of production. However, as has been shown in section 7.3, the size of the domestic market is so limited that a policy of this kind would have very little real effect. The 1984 imports, in terms of units, amounted to 109 units or 44 per cent of investment. Using the estimate of demand for 1988, a *full* (100 per cent) import substitution policy would add only another 175 units to the local suppliers. Hence, apart from being a blunt instrument, a tariff would also be of marginal importance for the local firms' endeavour to increase their scale of output. The limited size of the domestic market is recognized by the firms in question and all of them already consider the developed country market, especially the US market, as their *home* market. The main problem for these firms is therefore to increase sales in these markets. As in the case of the Swedish firm (and to some extent the UK firm) discussed in section 3.4, the firms in Taiwan have to face the fact that the local market is very small in relation to the minimum efficient scale of production. At some point, firms in this situation may have to spend very considerable resources in order to develop a marketing network and a reputation abroad.

7.7.2   *The Export–Import Bank*

Very little of the lending from the Bank goes to machine tool firms – only a few million dollars and then mainly from the Fixed Rate Relending Facility (FRRF). This sum of a few million dollars should be set in relationship with the total exports of machine tools of 177.5 million USD in 1981. Hence, the activities of the Bank are of marginal importance to the firms. Indeed, out of the six firms interviewed, only one said it used the FRRF facility and another one the pre-shipment financing scheme. The market leader suggested that the FRRF facility was used only rarely as it was no good for complex machine tools owing to a three-year limit on the loan. One of the smaller firms said that it never used any government scheme because of the complexity of the application process. It was also suggested that the customers had their own credit and did not need to borrow from the Taiwanese bank. All in all, the Bank seems more suited to the needs of shipbuilding and other projects where long lead times are the rule.

7.7.3 Risk capital

The part of the programme that lends risk capital to firms is by its nature selective. This means that, in principle, funds need not be spent on firms that do not require them or on firms that are in such a position that they are not judged to be able to make the strategic change.

Two of the firms discussed in section 7.5 receive loans from the Bank of Communications for the expansion of production of CNC lathes. Firm C in Figure 5.1 is receiving funds for the construction of a new plant, which will contain an FMS for the production of CNC lathes and machining centres. This firm is right now trying to consolidate its entry into the low-performance strategy and the provision of risk capital for the construction of the plant, as well as for developing design skills, appears to be a very good example of a constructive government policy based on the actual needs of a firm in the process of shifting strategy from engine lathes to CNC lathes.

Firm A in Figure 5.1 has been offered the services of the Bank but, as mentioned above, it has already consolidated its entry and was therefore, in 1983 at least, not in need of extra financing and has declined the offer.

Firm I, which has chosen the rather unusual strategy of going for a more complex lathe (four-axis as opposed to two-axis controlled) has also received funds from the Bank for the acquisition of high-quality machine tools, which are indispensable for the production of precision machine tools.

The general comment on this programme is that it is well suited to the actual needs that can arise as a firm switches its strategy.

7.7.4 R&D on automation

MIRL was specifically created to enhance the performance of the machine tool industry. The most important characteristic of MIRL is that it concentrates scarce design skills in one place. This allows MIRL to do R&D on a subcontracting basis for firms as well as to undertake large programmes such as the one that involves designing a CNC unit. The shared financing of the programmes between MIRL, the state and the firms also allows MIRL to absorb some of the risks in R&D. MIRL also educates firms and transfers technology to them.

In the case of the CNC lathe producing firms, the value of MIRL differs according to the position they are in. The two most advanced firms, if they continue with the low-performance strategy, would not require the services of MIRL as a shortage of designers is not their main problem. Firm C would still need to improve its design capabilities but no drastic change is required, so an improvement can be achieved by a maturation of the skills of the existing designers and by a marginal increase in the number of designers.

For firms of type F, G, H and I, which have a severe shortage of designers, MIRL could provide the important function of bridging some of the technological gap. Indeed, MIRL has recently designed two CNC lathes for such firms.

For the leading firms, MIRL could become an important instrument as and when they decide to change strategy to the overall cost leadership strategy, which means, *inter alia*, moving up-market to producing medium-performance CNC lathes. Such a change in strategy would, however, mean head-on competition with the Japanese firms. According to an interview with the leading firm, the firm is now planning such a change in strategy. It would then experience a number of problems, one of which would be a change in the design skills needed. Let us elaborate on this possibility.

In section 4.1, the need for design skills was discussed as well as the fact that this need has increased over the years with an increase in the rate of technical change. Among the Japanese firms pursuing the overall cost leadership strategy, the lowest number of design engineers is 115 and one firm employs substantially more than 200 design engineers. The leading Taiwanese firm has thirty-five design engineers. Of course, the number of designers needed is partly a function of the range of machines produced. Furthermore, several of the leading firms are doing pioneering work in design, including the CNC unit, which of course requires more resources than following or imitating. In spite of these qualifications, the difference in the *mass* of accumulated design skills means that the firm would need to spend nearly all of its R&D resources on developing its CNC lathe series, and would also need to keep these resources concentrated on CNC lathes in order to be able to prepare the subsequent generation of models. The firm has, however, also moved into the machining centre market and decided to shift some resources to the design of such NCMTs. This shift was clearly made at the expense of developing a complete series of CNC lathes in 1980–82 – as was mentioned in section 7.5(i), the third CNC lathe model was designed much later than the first two models in that series. The firm now has four models of machining centres. The first ones were low-performance models but it has now decided to try to move up-market with these products. The conflict between the design of CNC lathes and machining centres will of course be accentuated if the firm wishes to move into medium-performance machine tools with both products. A choice has to be made between the two.

Even if such a choice is made, the firm would need to increase its R&D expenditure very considerably, perhaps even to double it if it is to be able to compete successfully in the strategic group where firms follow the overall cost leadership strategy. Of course, this would not be easy. Financially it may create problems but the main constraint is the lack of core designers discussed in section 4.1. In such a situation, MIRL could well act as a temporary supplement to the firm's own design team. Indeed, it was reported that MIRL had entered into collaboration with firms A and C in Figure 5.1 for the development of a robot to be used for transferring components to and from the CNC lathes (*Metalworking, Engineering and Marketing*, 1983b, p. 84).

Another problem that should not be overlooked is that when firm A shifts strategy it will lose the cost advantage it had from designing low-performance CNC lathes. This means that the question of economies of scale becomes more important and that production volumes would need to be increased in order for the firm to remain competitive. Such a change in strategy would therefore imply an increase in total capital, which may by no means be easily available for such a high-risk venture. Indeed, access to risk capital at a time when a firm is about to change its strategy seems to be a very important obstacle for many firms (see the discussion on the Swedish firm in section 3.4). This leading Taiwanese firm recently went public in order to raise capital. It remains to be seen, though, if the private capital market can supply enough capital as and when the firm makes the shift in strategy. This firm can, however, always count upon the collaboration of the Bank of Communication whose programme was discussed above.

7.8 Summary and conclusions

This chapter has presented the case study of Taiwan. In section 7.2 I painted a broad picture of the historical development of the industrial sector and, in particular, the engineering sector. In contrast with the case of Argentina, the Taiwanese engineering sector is relatively small in relation to the manufacturing sector. The engineering sector is also of a different character than the Argentinian; in particular, electronics play an important role. The industry is also very open in terms of both exports and imports.

The development of the machine tool industry was discussed in section 7.3. The industry grew phenomenally in the 1970s, having played a marginal role in the 1960s. The growth of the industry was based mainly on exports, first to the Asian region but later to the developed countries. The export orientation was based on specialization in very simple machine tools. Price was the most important competitive tool. The implementation of a strategy based mainly on price competitiveness was aided by a stable real exchange rate. Other government policies were of little importance to the industry. Tariffs were and are low and, although import licensing exists, the import share of investment is very high.

As in the case of Argentina, the diffusion of CNC lathes only began at the end of the 1970s. The size of the market was a little more than 100 units per annum until 1984 when it increased to 250. Although it is expected to increase in the future, the local market is, and will continue to be, of limited importance to the firms producing CNC lathes.

The historical development and the present strategic position of six CNC lathe producers were dealt with in section 7.5. Very large differences exist between the firms as regards their resources.

I then outlined the policy of the government vis-à-vis the industry. I began by indicating the desire of the state to change the structure of the industry. A policy package covering 150 products, including CNC lathes, was recently designed. The implementation of this policy indicates a change in the attitude of the government towards the machine tool industry. The various elements of the policy package were outlined and evaluated from the point of view of the actual needs of the six firms described in section 7.5. I concluded that the trade policy of raising tariffs is both inefficient and of marginal use to the firms; that the activities of the Export–Import Bank are of marginal use to these firms and appear to be more suited to the needs of the shipbuilding industry; that the provision of risk capital to selected firms attempting to change their strategy is highly relevant to the needs of the industry; and that the MIRL can help firms experiencing an acute shortage of skills in a period of transition from one strategy to another. Finally, I mentioned the activities of the China External Development Trade Council, which helps firms overcome the obstacles caused by indivisibilities in the information-gathering process in foreign markets.

Note

1 Information on this and the next three firms dates from early 1983.

8
The Case of Korea

8.1 Growth and structure of the engineering industry

The Korean economy has experienced phenomenal growth in the past two decades. In fixed 1975 prices, GNP grew at a cumulative 8.2 per cent per annum between 1960 and 1981 (Nam, 1981; *Korea Annual*, 1982). Per capita GNP reached USD 1,880 in 1983 (Bank of Korea, 1984). The rate of growth in the manufacturing industry was even faster during this period – cumulative 15.7 per cent (*Korea Annual*, 1982). As a consequence, the share of manufacturing in GNP rose to 28 per cent in 1983 (Bank of Korea, 1984).

During the first three five-year plans, between 1962 and 1976, Korea followed an export-oriented growth based on the growth of labour-intensive light industries. The openness of the economy is reflected in Table 8.1, which shows imports and exports as a percentage of GNP. The Korean growth strategy could be labelled a consumer goods led strategy, much in line with that of Taiwan and Japan. The ratio between the value of production in the traditional consumer goods industries and in the engineering industry is shown in Table 8.2; the development greatly resembles that of Taiwan shown in Table 7.16.

Table 8.1 *The share of imports and exports in GNP in Korea, selected years 1960–1978 (%)*

Year	Exports	Imports
1960	2.7	12.3
1964	4.4	11.1
1968	11.2	25.9
1972	21.3	30.6
1976	34.9	40.6
1978	41.8	52.8

Source: Nam (1981), p. 190.

Table 8.2 *The ratio between consumer goods and engineering goods production in Korea, 1960, 1975 and 1981*

Year	Ratio
1960	5.7
1975	2.8
1981	1.8

Sources: 1960 and 1975: Edquist and Jacobsson (1982); 1981: Economic Planning Board, The Republic of Korea (1981).

The engineering sector's share of manufacturing value added rose only slightly from 10.7 per cent in 1960 to 16.3 per cent in 1975, while it grew to 25.3 per cent in 1982 (see Table 8.3). The structure of the engineering sector is, however, fairly similar to that of Taiwan in that the share of the non-electrical machinery industry is low and in that the electrical machinery industry is strong. Again, it is the production of electronics products that dominates the activities in this industry. In 1982, 66 per cent of the gross output of the electrical machinery industry (ISIC 383) was constituted by radio, television and communication equipment (Economic Planning Board, 1982).

Table 8.3 *Value added in the manufacturing industry and the share of the engineering sector in Korea, selected years 1960–1982*

Year	Engineering (current won m.)	Manufacturing (current won m.)	Share of engineering sector in manufacturing industry
1960	2,333	21,866	10.7
1963	6,299	61,534	10.2
1966	22,340	156,174	14.3
1969	59,802	426,041	14.0
1975	460,106	2,828,149	16.3
1979	2,224,823	9,207,982	24.2
1982	4,379,552	17,305,636	25.3

Source: Economic Planning Board (various).

Although important elements of import substitution exist in the Korean engineering industry (see below), the economy is fairly open as regards its investment needs. Imports as a percentage of apparent consumption were 47 per cent in 1980 (see Table 8.4). The figure is to be compared to the Argentinian figure of 12 per cent and the Taiwanese of 48 per cent.

Table 8.4 Production and trade in engineering products in Korea, 1980

Sector	Production		Exports	Imports	Apparent consumption[a]		Imports/ apparent consumption	Exports/ production	Self-sufficiency ratio[b]
	US$ m.	%	US$ m.	US$ m.	US$ m.	%	%	%	%
Fabricated Metal Products	1,462	14.5	747	210	925	8.4	22.7	51.1	77.3
Non-electrical Machinery	1,383	13.7	365	2,319	3,337	30.1	69.5	26.4	30.5
Electrical Machinery	4,121	40.9	1,918	1,606	3,809	34.4	42.2	46.5	57.8
Transport Equipment	3,103	30.8	1,150	1,050	3,003	27.1	35.0	37.1	65.0
Scientific and Professional Instruments, etc.	477	—	—	—	—	—	—	—	—
Total (excluding Scientific and Professional Instruments)	10,069	100.0	4,180	5,185	11,074	100.0	46.8	41.5	53.2

Sources: Economic Planning Board (1980); UNCTAD (1982b).
[a] Production minus exports plus imports.
[b] Production minus exports/apparent consumption.

8.2 The Korean machine tool industry

In 1982, ninety-one firms were registered as metalcutting machine tool makers in Korea (Economic Planning Board, 1982). However, most of these firms are very small and, in 1982, firms with more than 200 employees accounted for 57 per cent of the gross output.

The Korean machine tool industry originates from the period after the Second World War. Until the mid-1970s, however, the industry was fairly small and exports were insignificant. As can be seen in Table 8.5, in the second half of the 1970s the industry went through a period of explosive growth. Production rose from 5.2 million USD in 1971 to 178.4 million in 1984 and only 17 per cent of the value of production was exported in 1984. In contrast to the Taiwanese case, the fast growth in the production of machine tools was based on the rapidly growing domestic market. By 1979, Korea had developed into the tenth largest consumer of machine tools in the world (NMTBA, 1981/82). The import share of investment in machine tools is around 50 per cent. For lathes (see Table 8.6), the picture is fairly similar, although the export ratio is higher and the import ratio lower than for machine tools in general.

The price per imported unit is much higher than that of the price per exported unit (see Table 8.7). Like Taiwan, Korea has in this sense exploited the benefits of specializing its trade to the developed countries. A lower price per unit normally reflects weaker performance and/or lower quality. Data on the destinations of exports are found in Table 8.8. Whereas in 1977, when exports had just begun, the LDCs were the main export market, a shift to the developed countries, in particular the USA, had already taken place by 1979.

8.3 Government policy

The Korean government is deeply involved in restructuring the economy and fostering an engineering industry. Although the government introduced a series of laws to promote the heavy and chemical industries in the 1960s (Nam, 1981; Kim, 1982), it was only in the second half of the 1970s that the government especially emphasized the development of this sector.

A number of measures are, and have been, used to promote the sector. I shall briefly describe each of the main instruments. First, quantitative restrictions on imports have always played an important role in Korea (Nam, 1981). Kim (1983a) also emphasizes these restrictions in an excellent overview of the Korean promotion policies in the engineering industry. He shows that the Import Liberalization Rate[1], if measured by number of items, is consistently lower for the

Table 8.5 *Production and trade in machine tools in Korea, selected years 1971–1984[a]*

Year	Production[a] (US$ m.)	Exports (US$ m.)	Imports (US$ m.)	Apparent consumption[b] (US$ m.)	Imports/ apparent consumption (%)	Exports/ production (%)
1971	5.2	n.a.	n.a.	n.a.	n.a.	n.a.
1973	12.6	2.5	46.9	57.0	82.3	19.8
1977	73.7	1.6	152.6	224.7	67.9	2.2
1979	163.7	14.9	310.8	459.6	67.6	9.1
1981	178.0	30.6	130.2	277.6	46.9	17.2
1982	97.0	27.9	86.7	155.8	55.6	28.8
1983	144.4	21.5	139.7	262.6	53.2	14.9
1984	178.4[c]	29.8	135.0	283.6	47.6	16.7

Sources: Production data: 1971–81 – Economic Planning Board (various); 1982 – KMTMA (1983). 1983–84 – KMTMA (1985b). Machine tools are defined as ISIC 38231 and 38232.

Export and import data: 1977–84 excluding 1982 – Office of Customs Administration (various); 1982 – KMTMA (1983). Export and import data defined for 1973 as SITC 715 and for 1977–84 as BTN 8445.

[a] Includes production of parts and cast iron.
[b] Production minus exports plus imports.
[c] I have assumed an exchange rate of 840 won/dollar.

Table 8.6 *Production and trade in lathes in Korea, selected years 1974–1984*

Year	Production (US$ m.)	Exports (US$ m.)	Imports (US$ m.)	Apparent consumption[a] (US$ m.)	Imports/ apparent consumption (%)	Exports/ production (%)
1974	4.4	0.1	11.6	15.9	72.9	2.3
1977	23.4	0.3	18.2	41.3	44.1	1.3
1979	34.0	9.4	45.6	70.2	64.9	27.7
1981	34.0	21.0	20.4	33.4	61.1	61.8
1982	42.3	20.7	6.9	28.5	24.2	48.9
1983	45.0	11.3	8.6	42.3	20.3	25.1
1984	55.1	12.5	28.8	71.4	40.3	22.7

Sources: 1974–81: Economic Planning Board (various); Office of Customs Administration (various); 1982: KMTMA (1983); 1983 and 1984: KMTMA (1985b) for production data; Office of Customs Administration (various) for trade data.

[a] Production minus exports plus imports.

engineering industry than for industry as a whole – the ILR was 56 per cent in 1967/68 and 60.7 in 1983/4 for the engineering industry.

Secondly, the machinery industry has been favoured as far as both the volume of funds directed towards it is concerned, as well as the

Table 8.7 *Price per unit of lathes traded in Korea, selected years 1974–1984 (US$)*

Year	Exports	Imports	Export price as a % of import price
1974	26	6,353	0.4
1977	848	11,356	7.5
1979	6,465	23,817	27.1
1981	7,697	12,964	59.4
1982	6,567	13,651	48.1
1984	8,355	46,526	17.9

Source: Office of Customs Administration (various years).

Table 8.8 *Destination of exports of lathes from Korea, 1977, 1979, 1981 and 1984 (% of value)*

Year	North America	Europe	Japan	Australasia	Developing countries	Total
1977	15.7	—	14.9	—	69.4	100
1979	56.9	5.6	18.9	0.2	18.4	100
1981	54.4	6.6	17.6	5.8	15.6	100
1984	52.7	20.8	16.7	2.8	7.0	100

Source: Office of Customs Administration (various).

terms for using the funds. Thus, in the last decade, the engineering industry has consistently received more funds than other industries relative to its share of value added (Kim, 1983a, p. 16). Furthermore, more than 50 per cent of those funds consisted of preferential loans, that is loans involving large negative real interest rates (Kim, 1983, p. 23).

Thirdly, although not as important as quantitative restrictions, tariffs are widely used. Nam (1981) shows that the effective protection for domestic sales for the machinery industry rose from 44.2 per cent in 1968 to 47.4 per cent in 1978, whilst for the transport equipment industry the figures were 163.5 and 135.4 respectively.

Fourthly, a number of incentives aimed at promoting exports have also been introduced. For the manufacturing sector as a whole, these export incentives were greater than the incentives for sales on the domestic market. For the machinery and transport industries, however, the reverse was the case, at least in 1978 (Nam, 1981, p. 206). Other considerations need to be taken into account in order to get a more comprehensive picture of the operation of incentives. Corden (1981, p. 213) makes the point that: 'Perhaps the main feature was the absence [in Korea's growth experience] of restrictions on the availability of credit. Finance for exports could be obtained without difficulty.' Hence, the system of credits was geared towards export financing.

Kreuger (1981a, p. 215) adds another point of critical nature: '. . . in many industries, the right to sell in the domestic market at the higher effective protective rates . . . was generally reserved to those firms that were exporting. That is, protection in the home market was contingent upon being an exporter.'

Finally, as can be seen in Table 8.9, the fluctuations in the real exchange rate have not been that great, distinguishing the Korean from the Argentinian case, whilst being rather similar to Taiwan.

Table 8.9 *Fluctuations in the real exchange rate in Korea, 1967–1978*

Year	% change	Year	% change
1967	− 6.2	1973	.0
1968	− 1.7	1974	−16.5
1969	1.4	1975	− 1.0
1970	0.9	1976	6.0
1971	19.4	1977	0.0
1972	11.8	1978	2.4

Source: Nam (1981), table 2.

8.3.1 The machine tool industry

The performance of the machine tool industry in terms of growth is indeed remarkable, as was shown in section 8.2. In contrast to the Taiwanese case, the involvement of the Korean government in the machine tool industry appears to be considerable.

Subsectoral planning for the Korean machine tool industry began in 1977, with the formulation of two plans for the development of the machinery industry (Bendix *et al.*, 1978). Optimistic goals for the production and export of machine tools were set up and measures for supporting the development of the industry were outlined.

The government's interest in the machine tool industry was further underlined in the *1981 Basic Plan for the Advancement of the Machinery Industry*. The plan labels the metalworking machinery industry a *Leading Export Machinery Industry* (MCI, 1981, p. 21) and machine tools is 'one priority item to be assisted' (p. 6). The plan indicates a number of ways of assisting the machinery industry. One of these is protection. In the case of machine tools, the period of protection is specified as 'from the start of manufacturing to secure competitiveness' (MCI, 1981, p. 15). Another means of assisting the industry is 'improvement in loan conditions for, for example, R&D on specific

THE CASE OF KOREA 181

technologies and development of exportable products'. The plan also envisages great increases in exports. In the case of lathes, the actual exports of 14 million USD in 1980 were to be increased to 80 million USD in 1983 and 150 million USD in 1986 (MCI, 1981, p. 22).

I shall outline the central features of the 1977 plan and the present content of policy.

(i) The availability of long-term loans with low interest rates Whilst detailed data on the total allocations of funds to this industry are not available, two of the leading firms in the industry have both received large amounts of capital as part of the government's promotion of the machinery industry (see below).

(ii) Import prohibitions on items that can be produced locally All of the sixty-three items classified as machine tools at the CCCN 8-digit level are restricted (Kim, 1983b, p. 38). The noteworthy aspect of this policy is that it is the Korean Machine Tool Manufacturers' Association that has the power to decide which machine tools can be imported. If a product can be produced locally, an import licence is not given (interview). Of course, in an industry characterized by extreme product differentiation, the borderline between what can be locally produced and what needs to be imported can never be clear. In the case of CNC lathes in Korea, the present rule is that all CNC lathes below a certain size limit must be supplied by domestic producers. As this limit is set at a very large size, the vast majority of CNC lathe models cannot be imported. As is shown below (Table 8.10), the import share of investments in CNC lathes dropped to 41 per cent in 1981, which is when local production began to be significant. The implications for local machine tool users may be that they have to accept a less differentiated supply of machine tools, since Korean producers of CNC lathes produce only low-performance machine tools. In the OECD countries, the buyers of such machine tools are generally small and price-sensitive firms. Irrespective of the size of the CNC lathes, other firms generally demand higher-performance machine tools. However, in Korea, where the import restrictions apply to all CNC lathes below a certain size irrespective of their performance, all machine tool buyers have to settle for the low-performance Korean CNC lathes. It is of interest to note here that the situation is different in Taiwan where the body that decides on import licensing is neutral in relation to the machine tool industry. The higher share of imports in the total investment in CNC lathes in Taiwan could possibly be an effect of this institutional difference.

(iii) Financial assistance to firms buying locally made machine tools Kim (1983b, pp. 37, 38) explains:

On the demand side, the government is stimulating the development of the machine tool industry through credit and tax incentives. The government has put into place a buyers' credit system made up of the Procurement Fund for Locally-Produced Machinery for domestic buyers and Long-Term Export Credit Financing for foreign buyers. The government also provides special incentives to promote the use of domestically produced machine tools through enlarged tax exemptions and special depreciation allowances.

(iv) Tariffs According to Kim (1983b, p. 38) there is a 15 per cent nominal tariff rate on machine tools. However, given the quantitative restrictions, the tariff is of marginal importance.

According to Bendix et al. (1978) there is a condition for receiving incentives – the export of a certain percentage of production. According to sources at the Ministry of Commerce and Industry, exporters were looked upon more favourably when credits were allocated and the Ministry has regular meetings with the producers where their export performance is an important item on the agenda. Indeed, Korea is unique in that the Korea Machine Tool Manufacturers' Association publishes production and export data by firm every quarter.

Hence, it is clear that the machine tool industry is designated a priority industry by the government. The government is also willing to intervene deeply in the industry by providing low-interest capital as well as by applying import restrictions.

8.4 The Korean market for CNC lathes

The production, trade and apparent consumption of CNC lathes in Korea are shown in Table 8.10. Data on the share of CNC lathes in the total investment in lathes are also shown. Apart from the large demand for CNC lathes in 1974, the trend is similar to that of Argentina and Taiwan; it was only at the end of the 1970s that demand picked up. In terms of units, the demand was 215 and 248 units in 1983 and 1984 respectively. It is to be noted that the share of CNC lathes in total investment in lathes rose to nearly 40 per cent in 1984.

What of the future then? As in the cases of Argentina and Taiwan, I shall present a calculation indicating the possible magnitude of the future market in Korea. As I have recent data on the actual size of the market, I shall make an estimate for 1988.

As was shown in Table 8.6, the demand for lathes in Korea fluctuates somewhat. In this exercise, I assume that the demand for all

Table 8.10 *Production, trade and size of the local market for CNC lathes in Korea, selected years 1974–1984 (in units and value)*

Year	Production		Exports		Imports		Apparent consumption[a]		Imports/apparent consumption	Exports/production	Total sales of lathes	Share of CNC lathes in total investment in lathes (% of value)
	Units	Value (US$'000)	Units	Value (US$'000)	Units	Value (US$'000)	Units	Value (US$'000)	(% of units)	(% of units)	(US$ m.)	
1974	—	—	—	—	258	3,705	258	3,705	100	—	15.9	23.3
1977	—	—	—	—	16	1,054	16	1,054	100	—	41.3	2.6
1979	1	81	30	367	115	12,654	86	12,368	134[c]	—	70.2	17.6
1981	84	3,713	46	1,823	26	10,707	64	12,597	41	55	33.4	37.7
1982	222	11,250	138	7,105	18	1,885	102	6,030	18	62	28.5	21.2
1983	233	11,240	65	2,509	47	3,386	215	12,117	22	28	42.3	28.6
1984[b]	268	11,446	127	4,505	107	22,598	248	29,539	43	47	71.4	41.4

Sources: Production data: 1979 and 1981 – interview with the leading firm (data refer only to that firm's production); 1982 – data received from Ministry of Commerce and Industry; 1983 and 1984 – KMTMA (1985a). Trade data: Office of Customs Administration (various). Total sales of lathes: Table 8.6.
[a] Production minus exports plus imports.
[b] Assumed exchange rate of 840 won/dollar.
[c] Exceeds 100 because exports exceed production.

lathes will grow by 10 per cent annually from a figure that is the average between the actual markets in 1983 and 1984. I also assume that the share of CNC lathes in total investment in lathes will grow from the 1984 figure of 41.4 per cent to 50 per cent in 1988. Finally, I assume that the average price per unit will be 85,000 USD, which is the average of the years 1982–84. Given these assumptions, the annual demand for CNC lathes in Korea would grow to 490 units in 1988. This would mean a doubling of the market in comparison with 1984 (see Table 8.10).

Thus, although the size of the home market is and will be higher than in Argentina, the overall conclusion is that the potential home market will be rather limited in Korea, as in the other NICs.

8.5 The firms producing CNC lathes

In the first seven months of 1985, there were five firms in Korea producing CNC lathes. Total production amounted to 155 units. Two of the firms accounted for 86 per cent of production; these are firms B and D in Table 5.4 and Figure 5.1. I shall also briefly discuss a third firm, which mainly produces machining centres but which recently acquired the foremost German producer of large and custom-made CNC lathes.

(i) Firm B The firm started production from scratch in 1977 when a large engineering conglomerate 'following government directives to develop the heavy industries, . . . made massive investments into the machine tool industry' (Daewoo Heavy Industries Ltd, n.d., p. 26). In 1977, the production of machine tools was 239 units valued at 2.292 billion won (4.7 million USD). Production subsequently fluctuated somewhat, as can be seen in Table 8.11, but reached 10.6 billion won in 1983 (approximately 13 million USD).

Table 8.11 *Production of machine tools by firm B (in units and value)*

Year	Conventional lathes			CNC lathes			Other machine tools		Total
	Units	Value (won b.)	%	Units	Value (won b.)	%	Value (won b.)	%	Value (won b.)
1977	233	2,142	93	—	—	—	150	7	2,292
1978	924	7,086	91	—	—	—	686	9	7,772
1979	548	3,763	57	1	39	—	2,772	42	6,574
1980	227	1,431	37	9	210	5	2,244	58	3,885
1981	252	2,114	28	84	2,601	34	2,920	38	7,635
1982	281	2,532	22	183	6,052	52	3,038	26	11,622
1983	229	1,682	16	150	4,837	47	4,163	38	10,682

Source: Firm interview and KMTMA (1984).

As can be seen from Table 8.11, production was initiated with conventional engine lathes. The emphasis then shifted to CNC lathes and to some extent to other machine tools, which are predominantly milling and boring machines. In 1982 the firm planned to increase further the production of CNC lathes to 180 units in 1983, 220 in 1984, 330 in 1985, 390 in 1986 and 440 in 1987. It has, however, not been able to increase sales very much. In the first seven months of 1985, it sold 108 units, which on a yearly basis is equal to 185 units. The firm did, however, embark on the production of machining centres in 1984 and in the first seven months of 1985 it produced 35 units, which makes it the largest producer in Korea (KMTMA, 1985a). In contrast to the CNC lathes (see below), the production of machining centres is based mainly on an assembly agreement with a Japanese firm and is said to be limited to sales in the local market. The firm also has a licence for another type of machining centre from a German firm (Edquist and Jacobsson, 1985b).

The firm has invested very considerably in both equipment and skills. Altogether 44 million USD have been invested in production facilities, which include fifteen machining centres and twenty CNC lathes. The investment funds were allocated to the firm by the state on preferential terms, e.g. low nominal and negative real interest rate. As far as skills are concerned, in 1983 the firm had ninety engineers out of a total of 700 employees; sixty of these engineers are design engineers.

The firm spent 5 per cent of its sales on R&D in 1983. The firm began its development work by taking a design from a Japanese firm for the production of an engine lathe, but as early as January 1978 it had initiated its design development efforts on CNC lathes. Four reasons were given by the director for not choosing a design based on licence but rather going for the costly and uncertain development of own skills and designs:

- territorial restrictions on sales apply with licences
- only old designs are available
- royalties are expensive
- they are not allowed to adapt and develop the design.

In the design development of the firm, a very old design (see Chapter 5) was developed within twelve months. Considerable problems were, however, encountered with this machine in spite of its simplicity, as it was an adaptation of an engine lathe. During the subsequent years, six different models were developed but only the last can be said to be a commercial success; 140 units of this lathe were sold in 1982.

This CNC lathe is similar to a Japanese model but has, typically for the low-performance strategy, a weaker motor than the original. By 1982, the firm had therefore succeeded with only one model and this machine was a copy. The firm had however generated experience of design work. By 1982, the design engineers were all aged 28–35 years. At the end of 1982, the firm started to develop a series of four CNC lathes. The series, which was completed in 1984, is of its own conception and of medium performance. One of these was exhibited at the Machinery Fair in Seoul in October 1983 and the rest of the models have now been put on the market.

Together with firm A, this firm is the first NIC-based firm to have designed a complete series of CNC lathes. The director of the firm claims, however, that it does not yet have sufficient capabilities although he emphasizes the rate of increase in its capabilities and suggests that, in 1987–88, the firm will be able to supply CNC lathes with robots and other FMS features. It will be able to get some help with this from the engineering firms' central R&D laboratory, which has already copied a Japanese material handling robot.

An important reason why the firm has been relatively successful in its design efforts is that it has not fallen into the temptation of spreading its limited design people over too many products. In particular, the firm has chosen not to go for own designs for machining centres precisely because it cannot keep up with the international frontier for both products (Edquist and Jacobsson, 1985b). This is in sharp contrast to the Taiwanese firms' policy.

An interesting feature is that firm B had a flexible manufacturing cell installed in 1983 and it is planning to install a fully fledged flexible manufacturing system consisting of seven machining centres with a central control unit and an automatic transport system (Edquist and Jacobsson, 1985b).

The firm became export-oriented only two years after its creation. In 1979, it exported 291 engine lathes and sold 331 units in the local market. In 1981, the figures were 346 and 115 respectively. For CNC lathes, 60 units were exported in 1981 and 11 sold on the local market. In 1982, the figures were 139 and 19 respectively. Practically all exports have gone to the USA. Overall, the export ratio was approximately 70 per cent in 1982–83 but it has been reduced since then.

As would be expected, the company has not been profitable – only one year in the period 1979–82 was profitable. The official figures show only very small losses – 2–5 per cent of sales – but these figures probably do not reflect the true performance of the firm as it received very large, low-interest loans from the government (see section 8.3.1 and Chapter 9).

As for CNC lathes, the director said in 1983 that it would be profitable in 1985 with a production of 330 units per annum. An important reason for expecting to break even at this relatively low volume is that he claims that the firm gets the control units at a very good price. The director even claims that it gets the same price as the largest Japanese CNC lathe producer. Upon being asked the reason for this generosity he answered 'That is politics'. It can be added that Fanuc has been given permission to establish a production unit in the industrial park in the south of Korea where this firm is also located. The unit is a joint venture with a Korean firm (Kim, 1983b). As I noted above, the firm has not been able to expand its production according to plan.

As for marketing network, the engineering conglomerate has a worldwide marketing network. The company emphasizes the role of this network in keeping it in touch with the latest market developments. The machine tool firm also has six people in the USA for repair and maintenance, a group that needs to be strengthened according to the firm.

The present strategic position of the firm is that of having consolidated an entry into the low-performance strategy. The firm has also invested in creating design skills in a way that no other NIC-based firm has been found to do. This investment has laid part of the foundations for a change in strategy to one producing medium-performance CNC lathes. This change is now being attempted.

(ii) **Firm D** In contrast with firm B, this firm has long experience of machine tool production – twenty-five years. It is thus one of the oldest firms in this industry in Korea. In 1967 it introduced its first high-speed engine lathe. Engine lathes are still the firm's major product and it is the largest producer in Korea for that product. Other products of importance are milling machines and grinding machines, as well as CNC lathes. In Table 8.12 we can see the production in value of the firm for each product for the period 1 January – 31 July 1985.

Table 8.12 *Value of production by product in firm D, 1 January – 31 July 1985*

Product	Value (won m.)
Engine lathes	4,666
Milling machines	3,690
CNC milling machines	950
CNC lathes	1,162
Grinding machines	467
Total	10,935

Source: KMTMA (1985b).

The total sales of the firm were approximately 20 million USD in 1982, and it is thus one of the largest NIC-based firms producing CNC lathes. In total it has 800 employees. The firm's main market is the domestic market in Korea; its export ratio is only about 20 per cent.

The design development in the CNC lathe field began in 1977 when the Korea Advanced Institute of Science and Technology (KAIST) helped the firm to develop the basic design of its first CNC lathe (Lee, 1983). The model was not, however, very successful commercially and the firm developed its own model in 1981. It now has two CNC lathe designs and one machining centre design. The latter was put on the market in 1983 but, according to the Korea Machine Tool Manufacturers' Association (KMTMA), the firm had not sold any units of this machine in the first ten months of the year. On the other hand, 35 CNC lathes were sold in 1982, of which 20 were exported. The firm claimed in 1983 that it was producing 7–8 CNC lathes per month, but in 1985 production declined – only 26 units in the first seven months.

The firm had seventy engineers in 1983, thirty of whom were design engineers; ten of these had a background in electronics. The firm is therefore one of the largest NIC-based firms in terms of this criterion too. Apart from this asset, the firm has its own marketing network in the USA.

The present strategic position of the firm is somewhat unclear. The types of CNC lathes produced by the firm are small units of low–medium performance. In the export market, its customers are small and medium-sized firms, as in the case of all other NIC-based CNC lathe producers. In the local market, in contrast, its main customer is Hyundai Automobile, a very large firm. By October 1983, firm D had delivered 32 CNC lathes to Hyundai, which is planning to produce 300,000 automobiles per annum in collaboration with the Japanese firm Mitsubishi. These CNC lathes will operate in production lines producing, amongst other things, gears. Thus, in contrast to sales to small firms where the units are stand-alone CNC lathes, these machines operate in a completely different context. Normally, automobile firms buy their CNC lathes from firms pursuing the differentiation or focus strategy (see Chapter 3). Firm D has also developed a four-axis CNC lathe especially for Hyundai as well as a special material handling unit that allows for some degree of unmanned production. (The unit is fairly simple though and not of robotic type.)

(iii) **Firm J**[2] This firm has in the order of 1,500–1,800 employees and produced machine tools to a value of approximately 20 million USD in 1983. It produces mainly machining centres and engine lathes. In 1981 it developed a CNC lathe, and it produced 13 units in the first seven months of 1985 (KMTMA, 1985a). The firm appears to be fairly strong in its technological capabilities and it has, furthermore, installed a fully fledged flexible manufacturing system in its plant. Another interesting feature is that it is owned by Mr Moon, who is head of the World Unification Church. The firm has recently acquired two German firms that produce large and custom-made machining centres and lathes. Indeed, the German CNC lathe producer, Heyligenstadt, is often judged to be the foremost problem solver in this industry in the FRG. Thus, Korea has acquired this firm's capabilities through financial takeover.

8.6 Evaluating governmental policy

The Korean government has implemented a set of policies that are both general, e.g. tariffs and import restrictions, and firm- (and possibly product-) specific, e.g. allocation of investment funds. Given the importance of firm-specific policies in Korea, we would expect government policies to have affected firms very differently.

In the case of firm B, its rapid development reflects the Korean government's determination to compress the learning period involved in reaching the frontier. The investment in firm B – over 40 million USD – is very large for this industry. In comparison, the total Korean investment in tangible fixed assets in the machine tool industry was around 12 million USD in 1977 (Economic Planning Board, 1977). The investment in firm B was made with the help of a government loan with low interest, a preferential loan. Hence, the government absorbed all the risks in this investment. This applies particularly to the development of design skills in the CNC lathe field where it took the firm five failures before it learned how to design a saleable model.

Firm D also received government finance, although not as much as firm B. Firm sources claim that it received 5–10 million USD. Given the industry, this is still a large sum of money. Firm J also received government money, but I do not know how much. The importance of the loan for firm D is not clear but the firm began experimenting with CNC lathes at the time it received the loan. The firm also expanded its design team by 50 per cent in the period after the loan was received. A similar process of expansion took place in firm J.

Thus, in the case of firm B, the government policy was of a non-marginal character. It did not build on existing capabilities but enabled a new enterprise to accumulate skills rapidly. In the industry, this firm, as well as the government policy, is unique in this respect. In the case of firm D, the government built upon a fairly established firm and allocated it less funds.

The government also intervenes through the KMTMA by banning imports of machine tools produced in Korea. In the case of CNC lathes, this policy is of little use for firm B as it has the US market as its home market (see section 7.5.1). For firm D, the policy is apparently enabling the firm to establish itself as a supplier to the large Korean engineering firms. Normally, one would have expected a firm like Hyundai to buy its machine tools from a firm pursuing the differentiation or focus strategy. Although it is too early to say, the banning of imports could well enable this firm to become a national problem solver and avoid the low-performance strategy followed by all other NIC-based firms. This option is now also open to firm J, which can rely on the skills developed in the German firm it recently acquired. The costs to Korean society of this policy can, however, be high as was discussed in section 8.3.1 above. Furthermore, as these firms produce large numbers of conventional machine tools and firm D claims that the sales of these pay for the development of the CNC machines, the cost to Korean society could be high if these machine tools are priced above the international price.

Finally, the government measures relating to the demand side (see section 8.3.1) were thought of as being very important for firm B.

8.7 Summary and conclusions

This chapter presented the case study of Korea. In section 8.1, I painted a broad picture of the historical development of the manufacturing sector and the engineering sector in particular. As in the case of Taiwan, the engineering sector is relatively small in relation to manufcturing industry and is also very much concentrated in the electronics industry. The economy, as well as the engineering sector, is on the whole open to trade, although import restrictions are important tools in the government's policy.

The development of the machine tool industry, discussed in section 8.2, took place mainly in the second half of the 1970s when the government made it a strategic industry within the context of the development of the engineering industry. This industry grew in a phenomenal way after 1974, as did the demand for machine tools in Korea. The growth in the production of machine tools was mainly based on this growing demand in the local market. It was only in the early 1980s that exports began to become important. As in the case of Taiwan, the external market was initially the regional market but it was quickly switched to the developed countries. Although not as apparent as in the case of Taiwan, Korea has specialized in low-cost machine tools.

In section 8.3, I described the main government policies vis-à-vis industry. The Korean government is deeply involved in the restructuring of industry. It employs various means, the most important ones of which are control over investment flows and import controls, which are mainly quantitative restrictions. There is, furthermore, some evidence that the system of incentives is geared towards expansion on the export market. Finally, the real exchange rate has fluctuated less than in Argentina, but more than in Taiwan.

The local market for CNC lathes is in the order of 200 units per annum. As argued in section 8.4, this number is expected to increase to around 500 units by 1988, on the assumption of an annual growth rate of investment in lathes of 10 per cent and an increase in the share of CNC lathes in investment in all lathes to 50 per cent in 1988. The domestic market is then expected to develop to a size that is roughly half of one firm's output if it is to reach the minimum efficient scale of production for a firm following the overall cost leadership strategy.

Two firms, in the first seven months of 1985, accounted for 86 per cent of the production of CNC lathes (KMTMA, 1985a). The performance of these two firms was dealt with in section 8.5. The largest of these firms was established in 1977 as one part of the large-scale governmental effort to build a mechanical industry. The firm is unique in the sense that it was funded wholly by a preferential loan. On the basis of this loan, the firm rapidly accumulated skills and is today one of the leading NIC-based firms in this field. Naturally, no small firm operating with a profit constraint could have behaved in this way. The other firm has a longer history of machine tool production and has built up capabilities more gradually. This firm is different from other firms in the NICs in that it does not pursue a pure low-performance strategy. Although it exports to small and medium-sized developed country firms, it also sells to the local automobile giant Hyundai. Like firm B, it received funding from the government but on a smaller scale. A third firm accounted for 8 per cent of the production. This firm specializes in machining centres and engine lathes but recently bought the number one problem solver in Germany, Heyligenstadt.

Finally, in section 8.6, I evaluated the governmental policy. The machine tool industry became a strategic industry in the second half of the 1970s and has been favoured through various policies: credit, trade and the stimulation of demand. The two main elements in government policy – trade policies and credit policy – were discussed on the basis of the experience of firms B and D. The financial policy, being firm-, and possibly product-, specific, has been of particular importance to firm B. The decision to lend funds to the firm of course involved very high risks. On the other hand, if a country puts a high premium on rapid structural change, high risks at the level of the firm need to be taken. Import restrictions were not important for this firm (in the case of CNC lathes), as its export ratio is very high. The export orientation of this firm could possibly be explained by its total reliance on government funding. Firm D, which is more independent of the government, exports a smaller share of its production. For this firm, the import restrictions would seem to be of greater importance, not only in that they may generate a surplus in the sales of conventional machine tools, but also in that the firm may be choosing a strategy that is different from that followed by the rest of the NIC-based firms in the CNC lathe field. Firm J may now follow a clear differentiation strategy in the Korean market on the basis of the import restrictions and the skills of its German subsidiary.

Import restrictions for CNC lathes can be both inefficient socially and of marginal importance for firms pursuing a low-performance strategy. However, for a firm pursuing a strategy of differentiation (national problem solver), an initial ban on imports could enable the firm to establish relationships with large local firms. On the basis of intimate knowledge of the production process and needs of these firms, the machine tool firm could accumulate experience and skills and eventually be able to be internationally competitive. The costs to the society of being cut off from the world supply can however, be high and it is questionable if local demand is great enough for this strategy to be socially viable. Firm J's solution is interesting in that it may escape this market restriction through the takeover of the German firm.

Notes

1. $ILR = \dfrac{(T-R)}{T}$

    where
    - IRL = import liberalization rate
    - T = number of total items by CCCN 4 digit
    - R = number of restricted import items by CCCN 4 digit according to Ministry of Commerce and Industry's Annual Export and Import Notice.

2. This section is based on Edquist and Jacobsson (1985b). The firm is not included in Figure 5.1.

9

Government Policy

9.1 Introduction

In the last four chapters I have described and analysed the performance of the machine tool industries in Argentina, Taiwan and Korea. I have also discussed the contexts in which the firms constituting the industry have operated. The analysis was conducted against the background of the prior analysis of the structure and operation of the world industry producing CNC lathes. In this chapter, I shall discuss how government policy has influenced the performance of the firms and draw conclusions from my case studies on the efficiency of various types of government policies.

In section 9.2 I shall review some of the literature dealing with why and how a government should foster the development of an industry. I shall first outline the theoretical foundations of the infant industry argument. I shall then address the issue of whether a policy should be of general or of a selective character. Finally, I shall review the debate as to which policy instruments should be used.

After this review section, I proceed in section 9.3 to analyse the relationship between government policy and industry performance in Argentina, Korea and Taiwan. I shall first compare the performance of the machine tool industries in these countries as well as discuss the costs associated with their development. I shall then draw implications for government policy. The bulk of the section will try to answer the question of how government policy should be designed.

In section 9.4, I turn to discuss the industry performance in small developed countries and the implications for government policy. Some emphasis is given to the cases of Sweden and the UK.

Finally, in section 9.5, I summarize the chapter.

9.2 Survey of the main arguments

9.2.1 The infant industry argument

A number of reasons have been put forward over the past century in defence of state intervention to foster new industries. The infant industry argument is perhaps the most well-known argument. John Stuart Mill formulated this argument in an oft-quoted passage:

> The only case in which, on mere principles of political economy, protecting duties can be defensible, is when they are imposed temporarily (especially in a young and rising nation) in hopes of naturalizing a foreign industry, in itself perfectly suitable to the circumstances of the country. The superiority of one country over another in a branch of production often arises only from having begun it sooner. There may be no inherent advantage on one part, or disadvantage on the other, but only a present superiority of acquired skill and experience. A country which has this skill and experience yet to acquire, may in other respects be better adapted to the production than those which were earlier in the field: and besides, it is a just remark of Mr. Rae, that nothing has a greater tendency to promote improvements in any branch of production than its trial under a new set of conditions. But it cannot be expected that individuals should, at their own risk, or rather to their certain loss, introduce a new manufacture, and bear the burden of carrying it on until the producers have been educated up to the level of those with whom the processes are traditional. A protecting duty, continued for a reasonable time, might sometimes be the least inconvenient mode in which the nation can tax itself for the support of such an experiment. [Quoted in Corden, 1974, p. 248].

The argument for infant industry protection is an argument for a temporary intervention by the state. Corden (1974, p. 249) specifies two conditions for an argument to qualify as an infant industry argument:

- *time* must enter the argument in some essential way – it cannot rest solely on static economies of scale, whether internal or external;
- because it is an argument for intervention to alter the pattern of production, it will require some kind of distortion, imperfection or externality somewhere in the system.

Following Corden (1974) one can distinguish between internal dynamic economies and external dynamic economies as a basis for an infant industry argument.

Dynamic internal economies arise as a result of inhouse learning. As Corden (1974, p. 250) puts it: 'When factors of production are engaged in producing output in a particular year two products really result, visible current output, saleable currently on the market, and the invisible accumulation of experience and knowledge, in fact creation of human capital.'

As is noted frequently in the literature, the mere existence of dynamic internal economies does not justify intervention (Corden, 1974, 1980; Cooper, 1976; Little, Scitovsky and Scott, 1970). It is not unusual for firms to diversify into new product lines, be it a developing country based firm or a firm in the OECD countries. In fact, it would appear that it is most frequently the case that firms judge the future benefits of a new venture against the short-term costs and decide to pay the initial costs themselves.

Lack of information at the initial search phase (see section 3.2) would of course impede any potential entrepreneur even from seriously evaluating an area for business. Similarly, lack of, e.g., market information may impede a proper evaluation of a business idea. The correct procedure may then be to subsidize, e.g., market information. This is frequently done in many economies, including for example Taiwan (see section 7.6.5).

A more serious argument for fostering infant industries is imperfections in the capital market. As Corden explains (1974, p. 254): 'The potential infant may be unable to obtain finance to cover initial losses at a rate of interest which correctly indicates the social discount rate. There will be an under-investment, or failure to invest at all, in the creation of long term learning capital.'

There may be a number of reasons for such imperfections. The most important lie in different time-preferences for the individual and the state, as well as in different views on risks. It may simply be that the learning process is long and that the time required to break even is beyond the planning horizon of the individual entrepreneur whereas it is not beyond the horizon of the state. Thus, the social discount rate may be lower than the private (Cooper, 1976; Corden 1974, 1980).

In the context of the state attempting to induce a structural change of the economy, differential risk aversion between the individual firm and the state may be an important cause of under-investment in new activities. More specifically, in terms of the terminology of Chapter 3, there may be a failure to implement more radical strategic changes. As is well known in economics, it may well be rational for the state, or a very large firm for that matter, to invest simultaneously in, say, ten high-risk projects, being fairly certain that at least two or three of them will be successes. The same does not hold for an individual, smaller firm, which necessarily has to have a higher risk aversion than the state because its survival is at stake. Nor does it hold for banks lending money to such firms, which would need to increase the rate of interest to make up for the higher risks of lending to such firms. Corden concludes that 'There seems little doubt that, in spite of many qualifications, a valid, practically relevant infant industry argument for subsidization of new manufacturing industries resting on capital market imperfections can be made for many less-developed countries' (1974, p. 255). I would like to add that the argument may also have validity in developed countries.

Dynamic external economies may arise when the process of production or investment by a firm creates an invisible asset whose benefits in later years will go to another firm(s) without them paying the (full) price. Arrow (1962) presented the general case of external economies from learning by doing. Cooper (1976) discusses one particular case:

> Suppose that a firm in a less-developed country is about to invest in a new production line, and that the firm has a choice between using a local engineering design and construction company or a foreign one to carry out the investment phase. The local engineers have comparatively little experience. As far as the local investor is concerned, there are risks in contracting design and construction to them. They may produce a poor design, or they may be inefficient, so that the period of construction is extended, and unanticipated construction costs may be incurred. Consequently, the greater experience and longer track-record of the foreign engineers will put them in a strong competitive position, from the point of view of the local investing company. From a social point of view, there would be advantages if the local engineers were used. They would acquire experience and their efficiency would improve. In fact, since 'learning-by-doing' is a critically important source of experience and increased efficiency in design and construction work, there is really no other way in which they could improve their competitivity. So there would be social gains as well as gains to the local engineers themselves if they were employed by the local investor.
>
> However, these gains do not enter the private investor's calculus. He will not benefit from the improved efficiency of the local engineers, except possibly if he employs them a second time in the future – and even then, he would only benefit very partially. The gains in experience would benefit all other investment projects the local engineers undertook in the future. The investing company in this case would simply have to meet the costs incurred whilst the local engineers acquired experience.

Another source of external economies[1] has its origin in the Schumpeterian notion that it is the supplier who initiates economic change and the consumers who, if necessary, are educated by the producer (Arnold, 1983, p. 77). In the case of machine tools, Jones (1983) exemplifies such a source of external economies:

> ... the structure and competitiveness of the machine tool industry is very directly related to that of the engineering sector as a whole. Despite the growing degree of intra-industry trade and international specialization, there is still a strong link between a healthy domestic machine tool industry and a competitive engineering industry.
> ... countries who are dependent on importing the most advanced machine tools experience a certain delay in the diffusion of the latest machining technology. [Jones, 1983, p. 1]

The more generalized argument is that where new production processes are being used or when there are important elements of custom design in the production process, there are benefits to be reaped by the customers from having a local machinery supplier.

9.2.2 General versus selective state intervention

Given that there are widely acknowledged reasons for state intervention, it is logical to ask the question: should a protectionist system be uniform or made to measure? Corden answers the question by making a plea for uniform protection[2]:

> The essential idea of uniformity is that the same rates of protection be provided for all activities in manufacturing so that there is no discrimination other than that which comes naturally out of the price system. While a policy decision is made that manufacturing, or perhaps broad categories of manufacturing, be protected relative to other activities, no attempt is made to favor one type of activity within the protected area over another. Rather, the market ... decides on the pattern of industry. Thus, as far as possible, the principle of comparative advantage is applied. [Corden, 1980, p. 72]

Balassa agrees with this suggestion:

> ... the growth contribution of the manufacturing sector will be maximized – and the domestic resource cost of earning (saving) foreign exchange minimized – if all industries within the sector receive equal effective protection. Exceptions from this rule should be made only in cases when it is established that an industry generates substantially greater (lesser) external economies than the average. In so doing, one should avoid the use of 'tailor-made' tariffs benefiting a particular firm ... [Balassa, 1975, p. 374]

Balassa advocates a two-tier system of protection. Without providing any justification for the rates of protection chosen, he suggests a uniform effective protection at no more than 10–15 per cent for all manufacturing activities other than infant industries for which, 'exceptional cases aside, it does not appear likely that rates of effective protection more than double those for mature industries would be warranted on infant industry grounds' (p. 376). The principle of uniformity of incentives is thus upheld, but at two different levels.

Westphal (1981) makes a critique of the Balassa viewpoint and claims that:

> ... some countries – for example Korea – appear to have fostered the rapid achievement of international competitiveness by infant industries which on deliberately discriminatory grounds were initially granted whatever levels of effective protection were required to secure an adequate market for their output as well as a satisfactory rate of return on investment. [Westphal, 1981, pp. 12, 13]

Westphal then reformulates the 'prescriptions for infant industries'. He proposes keeping the Balassa recommendation of a uniform low level of protection for the manufacturing sector as a whole, 'though it does prescribe highly selective differential treatment to infant industries chosen on deliberately discriminating grounds' (p. 14). He advocates absolute protection for the infants; by this term he means

> ... whatever is necessary to secure an adequate market for the industry's output as well as a satisfactory rate of return on investment. Because it must reflect the efficiency of the industry, the precise level of protection can only be determined 'endogenously', for example basing protection on import quotas. [Westphal, 1981, p. 17]

Hence, Westphal departs from the Balassa view in two fundamental respects. He recognizes, first, that very high levels of protection may be needed at early stages in the infant's life and, secondly, that the same level of protection should not be applied to all infants because they operate under very different conditions and need to master different types of skills.

Westphal is less clear, however, about the reasons why greater selectivity results in a better performance. Whilst conceding that this is an area for further research, he puts forward some arguments of a more tentative character:

> Greater selectivity in import substitution undoubtedly accompanies

delaying the construction of initial plants until the market has grown to an appropriate size; it permits scarce investment resources to be concentrated in one or a few sectors at a time and thereby enables greater exploitation of economies of scale and of the linkages among closely interrelated activities. Greater selectivity equally allows the concentration of scarce entrepreneurial resources and technical talent and thereby avoids spreading the agents of technological change so thinly that no industry initially has the critical mass which may be necessary to initiate a sustained process of efficiency improvements through the acquisition of technological mastery. [Westphal, 1981, pp. 16, 17]

As regards the effect on the possibilities for a fast mastery of the technology, the arguments put foward are not fully convincing. In particular, because the infant industry argument implies that skills and capabilities are *created*, it is difficult to understand why, with an adequate educational system providing industry with *raw material*, a broad development of infant industries could not be generated. The problem for the decision makers in this case is rather whether or not society is ready to pay for the massive development of a large number of infants simultaneously. Indeed, given that infants can take a decade to mature, a practical reason for selective intervention is simply that society cannot find the means, or is not ready, to pay for the learning period involved in having a large number of infants simultaneously.

9.2.3 The choice of policy instrument

Once it is agreed that some kind of state intervention is justified, the question arises which instrument one should proceed to apply. Balassa (1975) discusses the benefits of tariffs compared with production subsidies and quantitative restrictions. He argues that tariffs are to be preferred both on account of the real world budgetary constraints and because the welfare losses from tariffs are presumed to be less important than the budget constraints of using production subsidies.

Balassa argues that tariffs are to be preferred to quantitative restrictions because

. . . while the direct effects of quantitative restrictions on the balance of payments can be easily ascertained, their indirect impact on the importation of inputs for the production of domestic substitutes and on export using inputs subject to restrictions is uncertain. Nor can their protective effects be established with confidence, thus making it difficult to evaluate their cost to the national economy. [Balassa, 1975, p. 37]

Balassa emphasizes in particular the problems associated with differentiated products where price comparisons are extremely difficult to make. These problems are accentuated as the economy becomes more developed, implying a greater demand for machinery and equipment – products that are often subject to great product differentiation. Tariffs, Balassa argues, are free from these problems because the buyer can decide whether or not to import a product by taking price, quality and performance into account as well as the tariff.

In contrast to suggesting *one* policy instrument only, Corden (1980) underlines the existence of a *hierarchy* of policies. The general aim is to get as high as possible in the hierarchy. Corden notes:

> . . . it is desirable to direct policies as closely as possible to the problem they are supposed to deal with. . . . Trade policies are not necessarily as close as other possible policies to the objective of fostering manufacturing. It is true that manufacturing may face its principal competition from imports, but this does not mean that tariffs or import quotas go to the heart of the issue. They are likely to create incidental and undesirable effects on the way – secondary (byproduct) distortions – which could be avoided by choosing instruments of policy more closely aimed at the objectives. [Corden, 1980, pp. 62–3]

Corden illustrates his argument by contrasting the effects of a tariff with the effects of a production subsidy. Whereas a tariff is equivalent to a sales tax on consumers, the revenue of which goes partly to the (benefiting) firm(s) and partly to the treasury, a direct subsidy avoids the problem of distortions in the pattern of consumption and still has the same effect as the tariff on the producer of the import-competing good.

Baldwin (1969) also criticizes tariff protection but on the basis that tariffs may not always be effective in obtaining the results desired for the *infant*, in contrast to the consumer. He argues: '. . . one cannot be sure that a temporary tariff will result in the optimum increase in production, or, indeed, in any increase at all in the production possibility curve' (1969, p. 296).

As an example, Baldwin takes the case of technological externalities and notes that, because there are costs associated with skill formation, firms may under-invest in skills that are mobile. He then argues:

> A protective duty is, however, no guarantee that individual entrepreneurs will undertake greater investments in acquiring technological knowledge. A duty raises the domestic price of a product and, from the viewpoint of the domestic industry as a whole, makes some

investment in knowledge more profitable. But the individual entrepreneur still faces the same externality problem. [Baldwin, 1969, p. 298]

Hence, protection of the whole industry is very much a second-best solution to the problem of skill formation or other sources of technological spill-overs. The point is generalized:

> ... when technological externalities exist, it is not enough merely to place taxes *or subsidies* on an industry's output ... specific taxes or subsidies directed towards particular types of inputs – for example in the infant industry case, towards research activities – may be necessary to achieve this goal [p. 300; emphasis added].

Thus, in some cases, not only tariffs but also production subsidies may fail to be the appropriate policy. Specific policies aimed at the problem to be dealt with are instead to be preferred.

9.3 Government policy and industry performance in Argentina, Korea and Taiwan

In Chapters 5, 6, 7 and 8 I described and analysed the development of the machine tool industries in general, and the firms producing CNC lathes in particular, in Argentina, Taiwan and Korea. Some emphasis was given to the role of government policies. The discussion in those chapters was conducted in the light of the changing competitive situation in the world market analysed in Chapters 3 and 4. In this section, I shall compare the performances of the industries in the three countries, discuss the costs associated with generating the industries and draw implications for government policies.

9.3.1 The benefits and costs of fostering the machine tool industry in the three NICs

(i) The performance of the machine tool industries The performance of the machine tool industries will be measured in four ways: (a) production performance; (b) export performance (being a very rough indicator of international competitiveness); (c) the technological basis of the industry for adjusting to the *electronic revolution*, which is indicated by the *mass* of design skills the firms have generated; and (d) production performance in terms of CNC lathes.

The production of machine tools in the three countries is compared in Table 9.1. It is striking that Argentina was well ahead of the other countries in the early 1970s. There is no doubt that, although the pricing system varies a lot between the countries, the Argentinian machine tool industry benefited from the early Argentinian demand for machine tools, which was a result of the Argentinian inward-oriented growth and early emphasis on the development of a domestic engineering industry. For example, in 1970, domestic demand for machine tools in Argentina was 47.8 million USD while the equivalent Taiwanese figure was only 13.7 million. The Argentinian machine tool industry was also undoubtedly helped by the early, high tariff rates on imported machine tools, whereas the Taiwanese government pursued a freer trade policy with only very limited tariff rates on machine tools. In the mid 1970s, however, both the Korean and the Taiwanese machine tool industries experienced phenomenal growth rates, and in 1979 both countries produced three times more machine tools than Argentina (in terms of value). By 1983, Korea and Taiwan produced machine tools to a value that was six or seven times as high as Argentina.

Table 9.1 *Production of machine tools in Argentina, Korea and Taiwan, selected years 1969–1983 (US$ m.)*

Year	Argentina	Korea	Taiwan
1969	17.6	5.1[a]	9.2
1971	22.2	5.2	12.8
1973	38.3	12.6	22.0
1974	n.a.	n.a.	33.2
1977	60.0	73.7	67.8
1979	62.0	163.7	189.1
1981	35.0[b]	178.0	242.3
1982	n.a.	144.4	n.a.
1983	28.2	178.4	211.0

Sources: Tables 6.4, 7.7 and 8.5; Korea, 1969 and 1974 – Economic Planning Board (1969, 1974).
[a] Metalworking machinery.
[b] Preliminary data.

In terms of exports, the Taiwanese machine tool industry is outstanding. It is highly successful in foreign markets, mainly the US market, whereas the industries of Argentina and Korea are predominantly oriented towards their domestic markets (see Table 9.2).

In terms of production and export performance combined, it is clear that Argentina has performed worse, in spite of its early start and lead in volume of production. Although the Korean and Taiwanese performances in terms of production have been broadly similar, Korean production is mainly sold on the protected home market, so one would

Table 9.2 *Exports of machine tools from Argentina, Korea and Taiwan, selected years 1969–1983 (in value and % of production)*

Year	Argentina		Korea		Taiwan	
	Value (US$ m.)	Export ratio (%)	Value (US$ m.)	Export ratio (%)	Value (US$ m.)	Export ratio (%)
1969	1.7	9.7	n.a.	n.a.	4.7	51.0
1971	3.9	17.6	n.a.	n.a.	6.3	49.2
1973	4.7	12.3	2.5	19.8	9.3	42.3
1974	6.7	n.a.	3.1	11.7	16.7	50.3
1977	8.5	14.2	1.6	2.2	47.4	69.9
1979	19.2	30.1	14.9	9.1	138.2	73.0
1981	20.0	57.1[a]	30.6	17.2	177.5	73.2
1982	n.a.	n.a.	21.5	14.9	n.a.	n.a.
1983	14.0	49.6[a]	29.8	16.7	135.4	64.0

Sources: Tables 6.4, 7.7 and 8.5; Korea, 1974 – Office of Customs Administration (1974).

[a] One needs to treat these figures with much caution and certainly should not believe that a trend change has taken place. Production was very low in these years, as was domestic demand for machine tools.

expect the price might be inflated. I would therefore suggest that the Taiwanese industry has been the most successful in terms of production and export performance combined.

In terms of lathes (see Tables 9.3 and 9.4), I would draw the same conclusions.

Table 9.3 *Production of lathes in Argentina and Taiwan, 1969–1979 (units)*

Year	Argentina	Taiwan
1969	1,552	3,228
1970	1,853	3,554
1971	2,003	4,569
1972	1,663	5,365
1973	1,973	6,490
1974	2,019	7,707
1975	1,623	7,933
1976	2,895	13,068
1977	n.a.	14,284
1978	2,660	17,649
1979	2,361	21,510

Sources: Tables 6.12 and 7.14.

Table 9.4 *Production and exports of lathes in Korea and Taiwan, selected years 1974–1984 (US$ m.)*

Year	Korea		Taiwan	
	Production	Exports	Production	Exports
1974	4.4	0.1	16.9	10.0
1977	23.4	0.3	28.7	20.8
1979	34.0	9.4	77.5	51.5
1981	34.0	21.0	86.0	67.0
1984	55.1	12.5	60.8	45.7

Sources: Tables 8.6 and 7.15.

As my indicator for the technological basis of the industries, I have taken the capability of adjusting successfully to the electronic revolution and of beginning to design and produce CNC lathes. The type of capability focused on has been design capability. The *mass* of design capability is, together with the ability to reap the benefits of economies of scale, the most important competitive factor in the industry. I studied the leading firms in the industries, which meant, practically speaking, all those producing CNC lathes. Detailed information was presented for five firms in Taiwan, two in Korea and one in Argentina. Only four of these firms were judged to have a reasonable chance of consolidating their entry into the CNC lathe industry. Two of these firms were Taiwanese and two Korean. These firms all have sales of 10–20 million USD and a design staff of thirty to sixty. The remaining firms, including the Argentinian firm, have a much lower volume of sales and only about five design engineers.

In other words, the Taiwanese and Korean industries show much

Table 9.5 *Production of CNC lathes in Argentina, Korea and Taiwan (units)*

Year	Argentina	Taiwan	Korea
1977	—	14	—
1978	—	40	—
1979	1	78	1
1980	2	106	9
1981	8	174	84
1982	7	131	222
1983	10	144	233
1984	n.a.	347	268

Sources: Argentina: Center on Transnational Economy (1984). Taiwan, 1977–81: data received from ITRI; 1982–84: data received from MIRL. Korea, 1979–81: interview with the leading firm (data refer only to that firm's production); 1982: data received from the Ministry of Commerce and Industry; 1983 and 1984: KMTMA (1985a).

greater technological capability than the Argentinian industry, where no producer is judged to be able to consolidate its entry into the production of CNC lathes, in the context of international competition, within the 1980s.

Finally, all three countries have embarked on the production of CNC lathes. Production data are shown in Table 9.5. Again, Argentina performs worst. Although Taiwan entered the industry several years earlier than Korea, the Korean firms have now caught up.

All in all, using the four indicators of performance, I would argue that Argentina has performed worst. Korea does not show the same performance as Taiwan in terms of the production and exports of machine tools and lathes, but it is equal in terms of the production of CNC lathes and it is as advanced in terms of design capabilities as Taiwan. I would therefore argue that, whereas historically Taiwan has performed better than Korea, Korea indicates at least as great an ability to adjust to the *electronic revolution* as Taiwan.

(ii) **The costs of fostering the machine tool industries** Allowing infant industries to be established and mature necessarily involves some costs to the society if these are not borne by the firms themselves. Like any other investment, investment in industrial capabilities thus involves both benefits and costs. It is very important to understand how the different policy instruments can also affect the cost side. As Bell *et al.* (1983) have noted, there is a great lack of studies that analyse both the costs and benefits of the establishment of infant industries. In this section I shall discuss the character of the costs as well as present some idea of the magnitude of the costs associated with the use of different policy instruments.

The Taiwanese industry initially developed with very little state intervention and can therefore be said to have involved very little social cost in its generation. The costs were borne by the firms themselves and the two leading firms reached the international frontier after twenty and twelve years respectively.

In the Korean case, government intervention was mainly in the form of credit policies and trade restrictions. It is possible to calculate roughly the magnitude of the subsidies received by what is now the leading producer of CNC lathes. It should be noted that the firm grew from scratch to being the largest producer of CNC lathes in the NICs in just five years. The firm received 44 million USD in what is called preferential credit from the government. Preferential credits are to be distinguished from general credits and from black market credit. The interest on preferential credit was, until 1982, lower than that on general credits, which, in turn, was substantially lower than that on black market credit. Thus, receiving credits, and especially preferential credits, involved a subsidized interest rate. The size of the subsidy depends on the type of credit received and the share of each credit form in the total credit market. It is not, however, possible to calculate the *equilibrium* interest rate, as the size of the black market for credits is not known. To calculate the size of the interest subsidy, I have therfore used the difference between the interest rate on preferential credit of the National Investment Fund, the largest source of preferential credits, and the interest rate for overdrafts on general loans. I then compared the size of the subsidy with the official profits of the firm in 1979–82 (see Table 9.6).[3]

A further source of subsidy is that associated with a cross-subsidy from home market sales to export sales. According to Westphal (1981), such arrangements are common in Korea. Data from the CNC lathe producer are available only for 1981 and for CNC lathes only. Assuming that the export price is equal to the international market price,[4] excluding transport and marketing costs, etc., the difference between the domestic market price and the export price can be said to be the same as the cross-subsidy. I then multiplied this sum by the number of CNC lathes sold in the domestic market. The result is seen in Table 9.6.

Table 9.6 *Calculation of subsidy of leading Korean firm in the CNC lathe industry*

Year	Official profit (US$ '000)	Interest rate subsidy (US$ '000)	Cross-subsidy (US$ '000)	Total subsidy (US$ '000)	Total sales (US$ '000)	Subsidy as % of sales
1979	−351	−1,000	n.a.	−1,351	13,582	−9.9
1980	−338	−750	n.a.	−1,088	5,886	−18.5
1981	+557	−300	−207	+50	10,907	+0.5
1982	−266[a]	—	n.a.	−266	15,496[a]	−1.7

[a] Assuming exchange rate of 750 won/dollar.

Two observations can be made from the table. First, it would appear that the firm has already reached a position close to break-even after only five years of operation. Secondly, the losses involved are relatively small in terms of dollars. In 1979, they amounted to about 1.35 million USD and in 1982 they amounted to around 266,000 USD (excluding cross-subsidy). The benefits, in contrast, are very substantial. Not only does the firm export the bulk of its production, but it also has a very substantial investment in skill formation, the benefits of which will be reaped in the years to come.[5]

Of course, one could always argue that the opportunity costs have been very high. This may apply to the skilled workers, but it does not apply to the engineers because they were mainly taken directly from the University, which has a very high output of engineers. The scarce factor, in social terms, was, and is, *experienced* design engineers and these have been *created* by the firm.

A systematic comparison with the Argentinian case is not possible owing to lack of data. A few points can, however, be made. As was mentioned in Chapter 6, the Argentinian machine tool industry operated behind high tariff barriers from the early 1960s to the end of the 1970s. Nominal rates of protection for lathes amounted to 150 per cent in 1960, and decreased to 65 per cent in 1976. Effective rates were probably also very high.

The direct financial costs to the users of machine tools in Argentina would be the price difference between locally made machine tools and equivalent products on the international market. Price comparisons of this kind are not easy to find, but one study claims that in 1971, i.e. after ten years of protection, the local price for engine lathes was 59 per cent higher than the international price while the nominal tariff in that year was 90 per cent (Amadeo *et al.*, 1982). Furthermore, and as was mentioned in Chapter 6, the degree of protection in the mid-1970s was very much to the satisfaction of the machine tool producers, which would also suggest that the local price was higher than the international price. Finally, the chief designer of the leading machine tool firm claimed in 1981 (interview) that it could never compete in the international market. Although this statement should be considered with care, it nevertheless suggests that local prices were still considerably higher than international prices.

Hence, the costs to the Argentinian users of machine tools appear to have been high, not least because the industry has been protected for nearly twenty years. Although it is not possible to specify the total cost, it is nevertheless illustrative that a mere 10 per cent price difference between the local prices and international prices would mean a yearly social cost equal to the *total* cost to Korea of fostering their main producer of CNC lathes in 1979 and in 1980 respectively.[6] Given the length of the protective period, there seems to be good reasons to claim that the cost to Argentinian society of fostering the firms that today have a production capacity in the lathe area has been far greater than the cost to Korean society of fostering their single firm that produces roughly as much as all the Argentinian firms.

To summarize the discussions in these two sections: I suggested that Taiwan's performance was superior to that of Korea, whose performance, in turn, was superior to that of Argentina. Korea's adjustment capacity to the *electronic revolution*, i.e. its performance in terms of design capacity generated as well as production of CNC lathes, is at least as good as that of Taiwan.

Concerning the social costs associated with generating these industries, I argued that it had been close to zero for Taiwan. In the case of Korea, data were available only from the leading firm in the area of CNC lathes, but the costs associated with generating this firm appear to be lower than the costs to Argentinian society of fostering their whole lathe-producing industry, although the value of production is roughly equal. The remaining lathe-producing firms in Korea have, however, been subject to a different policy mix from this leading firm. Whilst credit policies were the main instrument used for the leading firm, trade policies seem to have been the main instrument used for the majority of firms. This is certainly true for the second largest CNC lathe producer. To the extent that local prices are higher than international prices, the social costs may well prove to be higher than for generating the leading firm on account of the low export ratios of the industry as a whole.

In conclusion, although no exact data can be presented, I would suggest that the ratio between benefits and costs is the lowest for Argentina, medium for Korea, and highest for Taiwan.

9.3.2 Implications for government policy

Against this background of the benefit/cost ratios of the industrial policies pursued, I shall in this section discuss the empirical material in the light of the policy issues brought out in the review section. I shall first discuss the reasons for state intervention, and then how intervention should be undertaken.

(i) The reasons for state intervention A number of points can be made with respect to the question of why the government should intervene.

First, the existence of imperfections in the capital market was found to be a valid argument in favour of state intervention in section 9.2.1. Given that either this argument applies or that there are external economies, one may note that the application of the infant industry argument is not necessarily restricted to the first time an industrial activity is undertaken in an economy. Katz (1983a, p. 36) asks whether 'society should take upon itself to protect a second, or even a third round of indigenous learning in order to prevent the technological gap to widen?' In other words, when the technological frontier moves quickly, a formerly mature industry can again become an infant. In the case of lathes, it is clear that the technological gap involved in going from zero production, or perhaps the production of some other mechanical engineering goods, to the production of engine lathes is smaller than going from the production of engine lathes to CNC lathes. This is well illustrated in the case of Taiwan, where firms that clearly have become *grown-ups* in the production of engine lathes, including substantial exports to the USA, are now equally clearly infants in the production of the new technology – CNC lathes.

Secondly, given the theoretical foundations of the infant industry argument, one would expect the scale and length of the intervention to vary depending on the product as well as on the country considered. In the case of engine lathes, the Taiwanese experience amply demonstrates that the difficulties involved in growing up do not need to be substantial. The industry is clearly characterized by very low barriers to entry, as witnessed by the fragmented structure in all three countries. Indeed, several of the Taiwanese firms in Table 5.1 began producing lathes about 1970 and became very successful exporters in the second half of the 1970s. Because of the low barriers to entry, these firms were able to afford the initial costs involved in setting up the business without any government intervention. In the case of CNC lathes, however, the same firms are trying to finance their entry into CNC lathes internally through their profits on engine lathes, but the very large entry barriers means that a firm needs to be very large to succeed. It is indeed questionable that any of the firms involved can successfully enter in this way owing to the size of the technological gap. The risks involved for the individual firm, or lending institution, would also be very high. Thus, in the case of CNC lathes, an intervention of a very different scale and length would be required compared to engine lathes. In other words, an intervention based on the difference between private and social discount rates may be required.

Thirdly, given that there is a learning period involved in setting up the production of a new product, there are always external economies of the Arrow-Cooper type. These external economies can, however, be of considerable importance in the early phase of the product cycle. As will be argued in section 9.4 below, this is the case with FMM today and even more with larger system developments.

Fourthly, another source of external economies lies in the relationship between a healthy and advanced domestic machine tool industry and the competitiveness of the local engineering industry. The local machine tool industry can, if it is efficient and advanced, have a direct effect on the rate of diffusion of new technology in the local engineering industry. The machine tool firms have to sell and therefore educate local users about the advantages of the new technology. Given the relatively small markets of the NICs, it is to be expected that the developed country firms will have less incentive to spend resources, which indeed can be substantial, on educating customers in these economies. There is also a greater likelihood that local machine tool firms have lower fixed costs of supplying service and maintenance than do foreign firms. This argument would also apply to the teaching function.

There may be another benefit from having a local machine tool industry. Whilst most machine tools that are sold are standardized, there is also a demand for more custom-designed machine tools. As the design of these involves a great deal of communication between the producer and the buyer, there may be obvious benefits from procuring the machine tools locally. A case in point is the Taiwanese MIRL, which claims that it can supply custom-designed machine tools at less than half the international price. Another case could be that of firm D, in Korea, which has sold a large number of CNC lathes to an automobile producer in Korea. Some of these lathes are of a four-axis type, which is very advanced and particularly suited for very long series. Some of the CNC lathes also included custom-designed peripheral equipment.

There is thus ample evidence that state intervention could be justified in this industry. In particular I would emphasize capital market imperfections, but external economies are also present. Of course, it is not self-evident that the NIC governments should use their scarce resources to foster this particular industry instead of some other industry, but both the Korean and the Taiwanese governments have specified CNC lathes as a strategic product. This policy decision is taken as a starting point for the subsequent analysis, which deals with the issue of how to intervene. In this context it should be mentioned that, although the barriers to entry into the production of CNC lathes are far higher than those for engine lathes, they are nevertheless fairly low in comparison with other high-technology industries. For example, in telecommunications, which is another industry affected by microelectronics, the leading firms employ around 2,000 design engineers (Göransson, 1984), which is about ten times more than the leading Japanese firms producing CNC lathes. Thus, the opportunity costs involved in developing competitive firm-specific resources in the lathe producers should not be exaggerated.

(ii) How should governments foster the production of CNC lathes? Whereas the production of engine lathes involves overcoming very low barriers to entry, the successful international sale of CNC lathes involves overcoming very high barriers to entry. Indeed, in terms of similarity of resources required, the step from producing no lathes at all or from producing, say, a textile machine, to producing engine lathes is smaller than the step from producing engine lathes to CNC lathes. Although external economies may be present, as referred to in section 9.3.1, the main argument for state intervention here is the non-marginal changes in the barriers to entry in the industry. These arise either as firms start to produce CNC lathes and begin to pursue the low-performance strategy or when they shift from this strategy to the overall cost leadership strategy. For the NIC-based firms, the question is not one of a gradual advance into marginally stronger positions in terms of financial and technological capabilities. Radical changes are instead required. Design personnel and sales need to be doubled, and production and export marketing capacities need to be strengthened accordingly. Such radical changes in strategy certainly involve a great deal of risk-taking and there are good reasons for assuming that one cornerstone of the infant industry argument, namely that of imperfections in the capital market, applies. It would apply not only to firms beginning to pursue the low-performance strategy but also to firms shifting to the overall cost leadership strategy.

Given that there are good reasons for implementing state policies, the question arises whether these should be of a general or of a selective character.

Corden (1980) and Balassa (1975) advocate neutral protection – but neutral in relation to what? The basis for the infant industry argument is that it takes time to learn an industrial activity, so the reference points ought to be the learning time and the size of the initial technological gap. If we compare two infants, say machine tools or CNC lathes and the assembly of home electronics or engine lathes, we will find that both the length of the learning time and the gap are greater in the former activities than in the latter. Thus, a uniform protection constituting an average between the two would discriminate against the production of the more advanced products. I would therefore argue, with Westphal (1981), that intervention in different products needs to be tailor-made. The need for product-specific intervention is also advocated by Katz (1983a, p. 136): 'In view of the peculiar features of each situation, and the specificity of each maturative sequence, it is impossible, in our view, to speak of protection programmes that are unvarying from one country to another or from one branch of industry to another . . .'

Once one suggests that different products require different degrees as well as lengths of fostering, a further critique of Balassa's suggestion of a low uniform protection follows. The cost to the society of an investment in skills, like any other investment, is foregone current consumption. Therefore, the larger the gap to be bridged, the greater is the current consumption that must be forgone. However, societies may well exist that prefer to forgo present consumption for a long time period in order to achieve a radical structural change in their industry. The Korean example in CNC lathes is a case in point, where the firm entered directly into CNC lathe production rather than concentrating solely on the simpler engine lathes. It is therefore not correct for an outside economist to state that all manufacturing activities requiring more than, say, 30 per cent protection should not be established in a society. It all depends on the objective function of the society.

As regards differences between firms, my main criticism of the Corden–Balassa view of uniform effective protection as the principle of state intervention is that they are applying principles derived from static allocation theory to questions relating to infant industry problems. These are by definition a matter not of static analysis, but of the growth of resources over time. The principle is nevertheless applied to the infant industry problem. These authors presumably assume that the process of transition from infancy to maturity will be efficient if this principle is applied. It can, however, be efficient only if one assumes the same *need* for intervention in all firms. That is, firms must be perceived to behave in the same way and therefore have the same infancy problems. As is often the case in economics, the analysts would appear to have a *representative firm* in mind when making their policy prescriptions. Instead, I suggested in Chapter 3 that a *strategy model* of firm behaviour could be a more appropriate tool in analysing industries and firm behaviour. To the extent that this strategy model is the appropriate one, and this appears to be the case in the industry of CNC lathes, the application of a general policy for promoting an infant industry could involve a more costly transition period to the society than a firm-specific policy for two reasons.

First, the strategy model suggests that it is only when firms attempt a strategic change – be it entering into the industry or shifting strategic group – that the infant industry argument may apply. In the Taiwanese case, this is illustrated by the difference between the leading firm and the second largest firm. The leading firm is well consolidated in its entry into the low-performance strategy, so it cannot be said to be a high risk-taker as long as it continues to follow that strategy. On the other hand, the second largest producer of CNC lathes in Taiwan, which produces a very similar product, is attempting to consolidate its entry into the low-performance strategy. The firm is investing substantial amounts of money in new production facilities and in new skills and can well be classified as an infant and high risk-taker. Thus, although both firms produce a nearly identical product, only one of them can be said to be eligible for government support on the basis of the infant industry argument. A policy that is only product-specific, be it a tariff, a production subsidy or credit policy, is wasteful if a firm that does not need support would receive it.

Secondly, the strategy model suggests that firms can be very heterogeneous in terms of resources although they produce a functionally identical product. The heterogeneity applies also to firms within the broad groups defined above and is well illustrated in Table 5.4, which specified some important characteristics of eight NIC firms producing CNC lathes. The heterogeneity suggests that some firms stand a better chance than others of consolidating their entry. Again, in the Taiwanese case, there are at least six firms producing some kind of CNC lathe. Given that the break-even point is at the absolute lowest an average of 500 units per annum or more, not all the firms can be expected to succeed. Indeed, if they were to break even in, say, five years, it would mean a Taiwanese production of at least 2,400 units in 1990, or an annual growth rate of around 50 per cent. This is judged to be very unrealistic. If the government wants to use its resources in an efficient way, a policy instrument would need to be able to pick only those firms that stand the best chance of succeeding. Hence, some firms should be singled out for support as, and when, they attempt to move to a high-risk strategy. Of course, the same selection could be made with a uniform protection but then inefficient firms would also receive resources. Taken jointly, these two arguments would suggest that the social cost of transition can be higher when a uniform effective protection is applied, in contrast to a firm-specific intervention.

Having argued for product- and firm-specific policies, I shall now discuss the type of instruments to be used in the application of such policies.

A view commonly held by economists is that the maturation process of an infant can be conceptualized as a gradual reduction in the initially high costs of production to a point where the production costs are equal to, or below, the international price.

> The essential point stressed by infant-industry proponents since Hamilton . . . and List . . . first wrote on the subject is that production costs for newly established industries within a country are likely to be initially higher than for well-established foreign producers of the same line who have greater experience and higher skill levels. However, over a period of time new producers become 'educated to the level of those with whom the processes are traditional' . . .; and their cost curves decline. The infant industry argument states that during the temporary period when domestic costs in an industry are above the product's import price, a tariff is a socially desirable method of financing the investments in human resources needed to compete with foreign producers. [Baldwin, 1969, p. 296]

The Mill–Bastable test is a formalized version of this conceptualization. For example, Bell et al. (1983) envisage the learning period as in Figure 9.1, in which OA is the unit cost of production at the beginning of the infancy period, OF is the unit cost of importing the product, FAB is the undiscounted total cost of the infancy, and CBG is the undiscounted benefits from investment in the infant. With the proper discount applied, an evaluation of allocative efficiency can be made.

Whilst this view may be appropriate in an abstract discussion of the costs and benefits of infant industries, it is nevertheless inappropriate when it comes to discussing policy. Indeed, this view of the maturation process would appear to assume a market form of pure competition with a need to master some basic production skills. The actual decision-making process of the firm is, as was discussed in Chapter 3, much more complex and the maturation process will therefore have to be more complex. A blanket policy prescription, such as Balassa's (1975) recommendation of tariff protection, or a production subsidy, would not be appropriate.

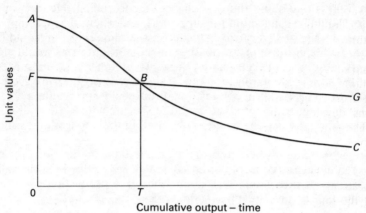

Figure 9.1 *The cost, benefit and duration of infancy*
Source: Bell et al. (1983).

It may well be the case that the firm can compete internationally pricewise but may simply not have access to a marketing network or the financial basis to build up its own organization. It may also be the case that the main risks for the firm lie in developing the minimum critical mass of design skills rather than in being an initially high-cost producer. In this case, which is relevant to CNC lathes, financial assistance to build up a large group of experienced design engineers is a much more appropriate policy instrument than a production subsidy. For example, the Argentinian government provides a 35 per cent export subsidy on the f.o.b. value, which is similar to a production subsidy. This policy is of little use when the firm cannot design a marketable product because it does not have enough design engineers. Hence, function-specific intervention may be called for. Of course, one could always argue that firms would spend the resources they received where their marginal productivity would be the highest. The main problem with this argument is that the amount of subsidy would be tied to the particular instrument used. For example, if the intervention was in the form of a production subsidy, the size of the subsidy received by firms may be linked to the number of units sold. A tariff would link the subsidy to the number of units sold to the domestic market. This would mean that firms would adapt their strategies in order to benefit from the intervention. Where a number of strategies are potentially of interest to the industry, only one would be chosen. A production subsidy in, say, Korea would not encourage the development of a national problem solver, for example, because the volume of output of such a firm is normally low. Thus, only a part of the potential industry would be developed.

Another important problem with a single instrument, especially tariffs and production subsidies, is that they are marginalistic in nature. If the fixed costs of establishing a marketing network need to be met, a production subsidy would not be of much help whereas a lump sum loan would. Hence, a blanket policy prescription would be limiting not only in that it would be concerned with only a part of the spectrum of potential strategies but also because of discontinuities in firms' developments.

The type of government policy would therefore need to be related to:

- The behaviour of the individual firms, i.e. whether or not they are attempting a strategic change;
- the long-term viability of a large number of firms pursuing the same strategy, as the market form of pure competition is not always the prevailing one; and
- the differences between the resources of the individual firm and the requirements for successful competition in the strategic group that the firm wants to enter.

The policy conclusions can therefore not be of a general nature but need to be custom-designed not only for the industry but also for the firm and the particular problems facing each firm.

This, of course, would not exclude the use of prohibitions on imports. For example, if it were deemed appropriate to foster a *national problem solver*, a temporary import prohibition might be the only way of creating the necessary marketing links between the supplier of machine tools and their users.

The suggestion that a firm- and product-specific policy may be the appropriate one is strengthened by Corden's (1980) proposition that the general aim is to get as high as possible in the hierarchy of policies and that it is desirable to direct policies as closely as possible to the problem they are supposed to deal with. Unless one assumes that all firms face the same problems at all points in time, this principle must necessarily mean that different policies should be used for different firms and products, even though Corden never, to my knowledge, draws this conclusion. Hence, once policies are chosen to be product-, firm- and function-specific, the choice of instruments has to be made according to the particular problems that are to be solved for each product/firm mix.

Of course, a number of qualifications must be added. First, the reasons for a firm-specific intervention may not apply to all situations. If the technological gap to be bridged is small and/or the dimensions of competition few (i.e. the strategy model does not apply), as in the case of engine lathes, a state policy of trying to pick winners may not always be able to grasp the importance of entrepreneurial talents in different firms and would risk creating a bias against potentially efficient firms. This point applies as well to cases where the strategy model is valid. There is always the risk that the best firms are not singled out for state support if firm-specific policies are implemented. It may well be so equal that protection will give a more assured outcome in this respect. The risks of not choosing the best firms would therefore have to be set against the advantages associated with the application of firm- and function-specific policies.

Secondly, the implementation of firm-specific policies would require a non-corrupt state in order to minimize the risks of choosing the wrong firms.

Thirdly, a competent bureaucracy would have to exist before the implementation of such policies. The size of the bureaucracy need not, however, be very great as only a very small number of the firms in an industry are likely to plan a strategic change at any given point in time.

(iii) The actual policies in Argentina, Korea and Taiwan What, then, have been the actual policies pursued by the governments in the three countries with respect to the production of CNC lathes? Table 9.7 summarizes the main variables controlled by the state that influence firm behaviour.

In the Korean case, firm B, which now is one of the largest producers of CNC lathes in the NICs and has the largest number of design engineers, was set up in 1977 as part of a very much larger scheme to develop a capital goods industry. The main elements in that scheme were quantitative restrictions on imports and a generous credit policy. In the case of CNC lathes, the R&D policy is also of some interest.

This particular firm received 44 million USD in credit, a very considerable sum considering that the three largest producers of CNC lathes in the NICs have total sales of 15–25 million USD. Although it is highly probable, it is not certain that the firm received the credit in order to produce CNC lathes, as opposed to other machine tools. The credit must be labelled as very high risk. Simultaneously, imports of CNC lathes in the same size class are banned.

The government intervention was thus product-specific in its trade policy and firm-specific in its credit policy. What is more, a preference for, or requirement of, a high export share was associated with receiving these benefits from the state. Although the bias was for the home market and the real exchange rate fluctuated somewhat, the firm chose a strategy of international specialization. The intervention from the state allowed the firm to operate without a short-term profit restriction, which permitted a notable accumulation of skills as well as a quick

Table 9.7 The main state-controlled variables influencing firm behaviour

Country	Trade policy	Credit policy	R&D policy	Character of the domestic market	Exchange rate fluctuations
Argentina	High general tariffs and export subsidies	—	—	Small and fluctuating	High
Taiwan	Neutral trade policy, liberal import controls	Firm- and product-specific, implemented in 1982. No particular policy prior to 1982	Firm-, product- and function-specific, useful in critical stages	Small but stable growth	Low
Korea	Product-specific import restrictions coupled with export preference	Firm- (and possibly product-) specific credits	Firm-, product- and function-specific, useful in critical stages	Large and fluctuating	Medium

move into the high-risk CNC lathe field. The firm is now well consolidated into the low-performance strategy and is shifting to the overall cost leadership strategy.

In terms of its R&D policies, the R&D institute KAIST, which is financed by the Korean government, was instrumental in helping the second largest producer in Korea to shift over to the production of CNC lathes by helping it with the basic design development of its first model. The R&D policy could be said to have been of some significance at that particular stage of the firm's development. The policy was product-, firm- and function-specific.

In Taiwan, the government changed its prior policy of non-intervention in the machine tool industry in the light of its ambition to develop a high-technology industry. In 1982, the government implemented a selective intervention programme that is firm-, product- and function-specific. This programme aims at absorbing some of the risks involved in entering into the production of, e.g., CNC lathes. The main instrument in the programme is credit policy, which draws on a fund of 250 million USD. The money is being lent, at slightly subsidized rates, to high-risk investments. An immediate effect of this new way of thinking was that the second largest CNC lathe producer decided to put a lot of effort into consolidating its entry into the low-performance strategy through expanding its production capacity and its skill base with a loan from this fund.

In terms of R&D policies, MIRL, which is partly financed by the government and has 120 mechanical engineers and 60 electronic engineers, plays a similar role to that of KAIST in Korea, although MIRL's section for machine tools is substantially larger. As part of its many activities, MIRL has designed two CNC lathes for smaller lathe producers that are just entering into the production of CNC lathes. For the leading firms, MIRL may likewise become important as and when these firms shift their strategy to the overall cost leadership strategy. This strategy requires, as was mentioned above, possibly doubling the number of design engineers of the largest firm in Taiwan. Financially, this may be a problem but the main problem would lie in a lack of experienced design engineers. In an actual shift of strategy, which the leading firm at least was planning in January 1983, MIRL could act as a temporary supplement to the firms' own design teams. Indeed, it was recently announced that MIRL had entered into collaboration with the two leading firms for the development of a robot to be used for transferring components to and from CNC lathes. Again, the policy is product-, firm- and function-specific.

In the Argentinian case, the government continued to use trade policy as its main instrument of intervention. The CNC lathe producer, very rationally, decided that it lacked the financial and technological basis for competing internationally in the area of CNC lathes. As it could not expect to receive risk capital from the government, the firm asked for, and received, 35 per cent tariff protection for CNC lathes. Receiving the protection meant that the firm saw no reason for continuing with a costly and high-risk design development in the CNC lathe area. It therefore discontinued its design development and obtained a licence from a Japanese firm. It is important to note that the receipt of tariff protection enabled the firm to switch from self-reliance in design development to a licensing philosophy as well as to continue with low-volume production. Thus, by using a tool of intervention that was not closely directed to the problem it was supposed to deal with, the state achieved a result that was totally opposite to the goal of creating an internationally competitive industry. On the other hand, the firm did manage to survive in the dismal Argentinian environment and through its licence agreement it is at least familiarizing itself with fairly modern product technology.

Hence, the Korean and Taiwanese policies do follow the conclusions for policy intervention drawn from a strategy model of firm behaviour: they allocate funds only to firms attempting a strategic change, i.e. they are firm-specific; they are product-specific; and the firms receive either risk capital or design help aimed directly at the problems of the infants, i.e. the instruments are function-specific. The Argentinian government in contrast, uses a policy that is not related to the infant's problem and is, indeed, counter-productive. The case is a good illustration of Corden's (1980) suggestion that it is desirable to direct policies as closely as possible to the problem they are supposed to deal with. It also supports Baldwin's (1969) claim that tariffs and production subsidies may not always be effective instruments to obtain the results desired for the infant industry.

The Korean and Taiwanese governments also apply trade restrictions to CNC lathes. Whilst the Taiwanese trade restrictions are marginal, consisting of a 10 per cent tariff planned in 1983 to be raised to 20 per cent, the Korean government prohibits imports of CNC lathes below a certain size, as mentioned above. Three observations can be made regarding trade restrictions. First, trade restrictions are general instruments in that they do not distinguish between different firms. They are therefore inefficient. Secondly, quantitative restrictions prevent domestic buyers of CNC lathes from benefiting from the international industry's role as a diffuser of innovations. Thirdly, the benefits accruing to the lathe producers from trade restrictions are marginal, in that the size of the domestic market is limited in relation to the minimum efficient scale of production. Hence, any benefits from cross-subsidies or from increasing the volume of production are marginal. The exception to this rule could be firm D in Korea, which in part follows a different strategy from the low-performance strategy. On the whole, however, from the point of view of the firm, trade restrictions cannot be seen as important, in relation to the other elements in government policy. From a social point of view, however, I judge that the severe import restrictions in the field of CNC lathes in Korea force the society to pay an unnecessarily high price for fostering the industry.

As far as future policy is concerned in Korea and Taiwan, it is clear that the high level of concentration of the international industry, in particular within the strategic group of the overall cost leadership strategy, can permit only one or two firms in smaller countries to become permanent members of the industry. It is also clear that the firms have a long way to go before they are equal members of the international industry. Substantial resources still need to be invested in these firms. Since the gap between these firms and the leading firms globally, which have a significant volume advantage, is still very large and the growth in the market for CNC lathes has slowed down, the pursuit of the overall cost leadership strategy is very risky indeed.

I therefore argue that any subsidies given to this industry should be concentrated on the one or two firms that show the greatest potential of succeeding. This does not, of course, exclude the possibility that firms receiving no subsidies can coexist with these two firms. For the two best firms I further argue that collaboration ought to take place in areas where economies of scale are important. This refers mainly to the procurement of the CNC unit, including the motor, and to marketing. Government support in the form of, say, risk capital could be made contingent upon such collaboration. It may also be worthwhile to think in terms of mergers.

In the case of Argentina, the still chaotic economic situation creates a high degree of uncertainty for all actors, including firm E. Rationally, the firm follows a *survival* strategy. An offensive industrial policy will simply have to wait until the macroeconomic situation shows some stability. If and when the government has achieved such a stability, it will have to decide whether or not it would be willing to put the required capital at this firm's disposal. What is clear, however, is that, for each year that passes, it will be increasingly difficult and costly to try to enter the overall cost leadership strategy. If that option is ruled out, a continued dependence on licence agreements and production for a protected local or regional market would seem to be the only alternative if the firm is to continue to produce CNC lathes.

9.4 Industry performance and implications for government policy in small developed countries

In this section, I shall discuss the options that are open to the European industry, with special emphasis on the cases of Sweden and the UK. In section 9.4.1 I shall analyse the size and character of the European market and try to specify the share of the three main strategic groups in the European market. The strategic groups are those analysed in Chapter 3 – namely, the overall cost leadership strategy, the focus strategy and the differentiation strategy. I shall then discuss the future strategic composition of the CNC lathe industry and attempt to specify how many firms Europe can aim to have within the main strategic groups in the future. In section 9.4.2 I shall limit myself to analysing the options open to Sweden and the UK. Although the UK is one order of magnitude larger than Sweden in terms of both production and demand for CNC lathes, both countries are still relatively small, within the CNC lathe industry, in comparison with the USA and Japan. I shall also indicate what type of firm strategy and government policy are required, in my opinion, in order to provide the foundation for a viable CNC lathe industry in these two countries.

9.4.1 The European industry

The total European market for CNC lathes is rather difficult to specify in a precise manner. In the four largest EEC countries and Sweden, the total market was in the order of 5,000 units in 1984. For the remaining European countries, one can only estimate the size of the market. In 1984, Japan exported about 700 units and the FRG exported approximately 370 units to the European countries apart from the four largest EEC countries and Sweden. If we add the units exported by the USA, France, Italy, Sweden and the UK, the total European market may have been in the order of 6,300–6,500 units. This is of the same order of magnitude as the markets in USA and in Japan. A very important difference, however, is that the internal European market is not as integrated as those of Japan and USA. Nevertheless, the 'home market' of Europe is as large as that of its main competitors.

The market shares of the three strategic groups mentioned above can be roughly estimated on the basis of Swedish, German and US data. In the Swedish case, on the basis of interviews with the leading suppliers of CNC lathes, it can be estimated with some precision that in twelve months in 1984–85 firms following the overall cost leadership strategy accounted for around 44 per cent of the market in units; firms following the focus strategy accounted for 31 per cent; and firms following the differentiation strategy accounted for the remaining 25 per cent. As is argued in Edquist and Jacobsson (1986), Sweden has probably progressed further than other European countries as regards the diffusion of FMMs, so we cannot directly transfer these results to other countries.

In the case of the USA, I had to use a less precise method. I equated small CNC lathes (less than 18kW motor) with the CNC lathes produced by firms folowing the overall cost leadership strategy; medium CNC lathes (18–37kW) with the machines produced by firms following the focus strategy; and large CNC lathes (more than 37kW) with the machines produced by firms following the differentiation strategy. Using this method for the USA in 1983, 75 per cent of the horizontal CNC lathes sold were sold by firms following the overall cost leadership strategy, 19 per cent belonged to the focus strategy, and 6 per cent to the differentiation strategy.

Finally, in the case of the FRG, the total market for CNC lathes amounted to 1,661 units in 1984. About 600 of these were imported from Japan and other countries for which the average price per unit was less than 72,000 USD. This was the average price of 'small' CNC lathes in the USA in 1983. In addition, some German-made CNC lathes are of low performance and if we add these to the 600 we come to a figure of approximately 800 units sold in Germany in the market segment belonging to firms following the overall cost leadership strategy. This represents slightly less than 50 per cent of the market.

The US market thus appears to be more strongly dominated than the European by firms following the overall cost leadership strategy. The reasons for this are unclear, but an associated feature is surely the much higher market penetration by Japanese firms in the USA than in Europe. In value terms, Japan accounted for 43 per cent of the US apparent consumption of CNC lathes in 1984, and in units their market share was as high as 61 per cent (elaboration on NMTBA, 1985/86, and trade data supplied by NMTBA). The highest equivalent figure in value for the main European countries was 28 per cent in the case of the UK (see Table 3.18). A second reason is that, although the strength of the motor is an important indicator of the lathes produced within each strategic group, it may well be that some CNC lathes with a weaker motor are equipped with some kind of customized material handling unit and should therefore be placed within the strategic group of differentiation.

It seems reasonable to assume that around 50 per cent of the European market for CNC lathes, in units, consists of smaller, standardized units of low–medium performance. On the basis of the Swedish figures, I would suggest that around 30 per cent of the market belongs to firms following the focus strategy. The remaining 20 per cent belongs to firms in the differentiation group and to firms producing vertical lathes. According to NMTBA (1984/85), 13 per cent of the

US market in 1983 consisted of vertical lathes. To the extent that this ratio also applies to Europe, this would leave 7 per cent to the differentiation group.

Given an annual market for CNC lathes in Europe of about 6,400 units, 3,200 would therefore be units sold mainly by firms following the overall cost leadership strategy; 1,900 would be units sold mainly by firms following the focus strategy; 450 units would be sold mainly by firms following the differentiation strategy; whilst 850 lathes would be of the vertical type.

This division between the various strategic groups refers to 1984. As was argued in section 3.4, we shall see a merger between the differentiation strategy (*inter alia* CNC lathes being equipped with automatic material handling units) and the focus strategy. Of course, the differentiation strategy will have a different content in the future. The firms in this group will probably supply systems of greater complexity than is the case today. What the merger means, however, is that the overall market for the firms within these groups will merge in the years to come. As a result not only will the firms within the present strategic group of differentiation face increasing competition from the firms in the present focus strategy, but the market in the future for the new, merged, strategic group would be around 2,350 units.

The question then follows: how many firms within each strategic group can be justified in Europe? A first point to make is that the lack of appropriate adjustment by US firms has led to a catastrophic decline in the US CNC lathe industry. The US share of world output of CNC lathes, in units, dropped from 36.2 per cent in 1975 to 6.7 per cent in 1984. This decline was documented in Chapter 3, where it was also shown that the European industry's share of world production, in units, of CNC lathe dropped from 35.5 per cent in 1978 to 21 per cent in 1984. There was also a reduction in the European industry's share of global output if we measure in terms of value, although the reduction was not as dramatic. In spite of the better performance of the European firms, the industry is still under great pressure from Japan. Only a part of the European industry has managed to adjust in a satisfactory way. To the extent that the remaining industry is not going to be able to rely on governments subsidizing unviable strategies eternally, a considerable adjustment still remains to be made. How could such an adjustment be designed? Let us first look at the overall cost leadership strategic group.

The size of the European 'home market' means that there is scope for one or two European firms to establish themselves as proper members of the overall cost leadership strategic group. In Chapter 4, I estimated that the minimum efficient scale of production could approach 1,000 units annually. With a price premium given to local producers, a low tariff and the higher transport costs for the Japanese firms, the break-even point may be reduced to 500–700 units annually. In 1984, Japan exported close on 1,800 CNC lathes to Europe. Assuming that almost all of these were produced by firms following the overall cost leadership strategy and that the present level of imports will continue the number of European firms that could follow this strategy would be two firms.

In Chapter 3 I described one Italian firm that follows this strategy, but at a yearly production volume of only 350 units. This firm should, however, be able to develop into a much larger supplier of low–medium-performance CNC lathes (in terms of motor strength) to the European market. Given that the German market is the single largest market in Europe, it would seem reasonable that a German firm should be the second entrant. Indeed, one German firm, Gildemeister, already produces simple CNC lathes in a profitable way in one of its divisions. This division could be one candidate for the second European firm in this strategic group.

Entering into this strategic group at a time when the growth in world production is very modest and when the leading firms are well consolidated in the group is, however, a risky venture. As the Boston Consulting Group puts it:

> Entering a volume business at a stage when the leading supplier has already gained a significant volume advantage and/or the market growth is likely to slow down is very risky unless there are clear indications that the leadership will falter. [Boston Consulting Group, 1985, p. 31]

Given the very good financial status of the leading Japanese firms (see Tables 3.10 and 3.17), there is really no basis for hoping that their performance will falter. Thus, the conclusion is that it will be very risky for a European firm to attempt to enter this strategic group. The risks would, of course, be greatest for firms with a small domestic market. A further risk-enhancing factor to take into account is, of course, that firms in the NICs, as we have seen in Chapters 7 and 8, are entering into this strategic group and can be expected to expand their sales in both the European and the US markets. These risks, which are considerable, have however to be taken if Europe is not to leave the largest segment of the market to the Asian nations.

As a slight digression, we can point to the example of the Swedish firm ASEA, which produces robots – an area closely related to CNC machine tools.[7] ASEA is an example of a firm in a very small country that has been able to become one of the largest firms of its kind in the world. ASEA is one of the world's ten leading electrical and electronics enterprises with invoiced sales of 4 billion USD in 1984. ASEA began producing robots in 1972 mainly to replace robots that they had bought in from other companies and that were, in their opinion, too rudimentary. At this time, robot production was at a very early stage. ASEA produced 25 units in 1975, a figure that rose to 2,000 in 1985. Two crucial objectives were identified by the firm:

- maintaining and enhancing technological leadership;
- the ability to reap the benefits of economies of scale through volume production in order to meet price competition (Barnevik, 1985, p. 6).

We can add that this presumably required a broad range of robots covering a number of applications. Today they cover about twenty applications (Sigurdson, 1985, p. 6).

In other words, ASEA emphasized both technological development *and* volume production as two basic elements of its strategy. Massive investments were, however, needed to implement this strategy. Whilst their sale in 1980 amounted to 20 million USD, they invested 100 million USD in the period 1981–84 – of this, only 20 million went to investments in machinery and equipment; the rest went to R&D and marketing. The investment in marketing facilities (the so-called Robot Centre) was particularly emphasized, accounting for approximately 40 per cent of the 100 million USD. This investment has paid off: in 1984 ASEA had a market share of 30 per cent in Europe, 10 per cent in the USA and 5 per cent in Japan (Sigurdson, 1985, p. 6).

Thus, in contrast to the European CNC lathe industry, ASEA has in the field of robots implemented a strategy that strongly resembles the overall cost leadership strategy. This was done in a country with an extremely limited local market, and it shows that firms in smaller countries can very successfully follow strategies that involve volume production. However, one crucial difference between the CNC lathe industry and this robot example is that ASEA has an *internal* capital market that can operate differently from the national capital market. ASEA is large enough and had the will to take the initial losses, which evidently were significant, and wait for the benefits, which are now beginning to come.

Returning to the case of CNC lathes, let us proceed to the new, merging, strategic group of focus/differentiation. The two leading firms in this group, which were described in Chapter 3, produce 500–600 units per year. Assuming that the break-even point lies around 500 units annually, there would be room for five firms of this type in Europe. Two to three firms in this strategic group have already reached this annual volume. There is thus room for another two or, at most, three firms to develop into fully fledged members of the merged strategic group. Two of these could well be the Swedish and UK firms described in Chapter 3.

Of course, this rough analysis assumes that governments or other institutions do not continue to subsidize machine tool firms that pursue *non-viable* strategies. If they do, there will be too many firms for the European industry to be viable in the long run. A possible solution would be for each government in the main machine tool producing countries to agree to subsidize only one firm each in these two strategic groups. For example, the Italian and German governments could foster one firm in the overall cost leadership group whilst the French, UK and Swedish governments, say, agreed to foster one firm each in the new merged group of focus/differentiation. A joint industrial policy by the European governments would, however, require a significant change in the thinking of these governments.

9.4.2 Sweden and the UK

In section 3.4 I referred to the problems of higher risks and greater uncertainty that firms in countries with a smaller domestic market face at times when attempting market expansion. This argument is put in general terms by Ergas:

> Firms in smaller states . . . face greater uncertainties in their search for market expansion. Both exporting and foreign direct investment involve additional risks and costs arising from fluctuations in exchange rates and competitiveness, overt and covert protectionism, the unfair competition of larger states on third markets, and the persistent differences between countries in technical standards, product regulations and demand patterns. For smaller states to succeed, they must consequently try harder . . . [Ergas, 1984, p. 7]

Thus, everything else being equal, firms in a small economy will need to take higher costs and greater risks or possess better information than their competitors with larger domestic markets, in order to perform equally well.

I also argued that the problems of firms in smaller economies are particularly acute at the time when a product is maturing or, in other words, when the product passes from the introductory phase of its S-curve into the growth phase. It is at this point that the firm needs to change its strategic orientation and take all the costs and risks of basing its expansion almost exclusively on foreign markets. In order to do this, a firm clearly needs to have access to very substantial amounts of capital, which by all criteria will be very high risk capital. It is by no means self-evident that the normal capital market can ensure a correct allocation of resources in these circumstances. The reasons for this were elaborated on in section 9.2.1. If the capital market does not allow firms to take the necessary risks associated with a strategic change, firms will opt for strategies that are associated with low rates of growth of output and employment. At least in the Swedish case, with the present budget and balance of payment deficits, such a choice of strategy would be socially undesirable in that it would not rectify the problem of an industrial sector that is too small. This effect was clearly illustrated in section 3.4, which showed that Sweden's world market share, in value, of CNC lathe production dropped from 4.63 per cent in 1975 to 1.39 per cent in 1984. It is worth mentioning here that production in Korea amounted to 0.8 per cent of world production in 1984. In terms of units, Korea, with a production of 268 in 1984, is already ahead of Sweden, which has a production of about 200 units. The same applies to Taiwan, which produced 347 units in 1984. In the UK, the drop was less dramatic – from 3.75 per cent in 1975 to 2.65 per cent in 1984.

Furthermore, in this particular industry, although the private profitability is not impressive, the social profitability can be much higher. I mentioned in section 9.2.1 that there are reasons to believe that there are positive external economies associated with having a well-functioning machine tool industry. In other words, there may be a positive relationship between the performance of the machine tool industry and the performance of the engineering industry as a whole. This is particularly so where elements of custom design are part of the machine tool or the system or when new products are introduced. The very rapid diffusion of CNC lathes in Sweden in the early and mid 1970s was probably related to the innovating role of the leading Swedish firm at that time. The same applies to FMMs today and will apply to an even greater extent to larger unmanned systems in the future. Indeed, a very close relationship exists between some of the leading edge user firms in Sweden and the CNC lathe producer in terms of the latter's product development. It is common in Swedish engineering firms to hear that a mutually beneficial relationship exists between them and the machine tool firm. It is thus without doubt true that the social profitability of this firm, and probably of other firms in the industry, exceeds its private profitability.

Given that there are good reasons for advocating some kind of government intervention in the industry, what objectives should it have? In other words, which strategic group should the government policy induce or help the firms to enter? Furthermore, which tools should the government use?

In Chapter 3 I argued strongly that the present strategic orientation of the leading Swedish and UK firms will not be viable in the future.

Essentially, the argument is based on the fact that the share of standard elements in the systems sold is already very high and will become higher in the future as the software becomes increasingly standardized. This standardization means that firms that can reap the benefits of economies of scale on the standard elements, in particular the CNC lathe, can and will increasingly be able to outcompete smaller firms on the basis of lower price. I also showed that the larger firms following the focus strategy are now rapidly accumulating experience as regards the sales of systems. Thus, the firms selling systems today will face growing competition in the future when an increasingly important element in the competition will be the price of the system. I therefore concluded that the leading Swedish and UK firms will need to increase their volume of output of standard CNC lathes in parallel with selling systems.

An alternative would be for the firms to stop producing CNC lathes and concentrate on selling systems alone. There are three drawbacks to this alternative, however. First, access to high-performance CNC lathes would be very limited because the producers of such lathes compete in the same market as the Swedish and the UK firms. Secondly, the advantage of larger firms would still prevail as regards the cost of the standard elements. Thirdly, from a national point of view this option would result in a reduction in industrial output and employment. I would therefore suggest that this alternative is neither viable nor desirable.

Sciberras and Payne (1985) also argue that the UK industry should develop its position in the CNC 'general purpose segment'. The means for improving their position are joint ventures between UK and Japanese firms, which include comprehensive technology transfer agreements. Sciberras and Payne argue:

> Specific skills and expertise, as well as particular product and manufacturing engineering technology, need to be identified and negotiated in exchange for the benefits of easier entry into the European market the Japanese firms will obtain from a joint venture arrangement in the UK machine tool industry [Sciberras and Payne, 1985, p. 86]

Essentially, their argment in favour of this solution is a perceived lack of management skills in the UK in the fields of strategic planning, production engineering and training and deployment. Access to knowledge as regards the CNC unit through the Japanese firm is also listed as a potential benefit of such an arrangement. Government assistance in the pre-negotiation, the negotiation and the implementation phases is seen as necessary for the UK firms to get a good deal.

Sciberras and Payne contrast this model of rationalization of the industry with several other models that involve reaching international competitiveness in the 'general purpose segment'. One of these models is a full merger of UK firms to maximize volumes of major product lines. Three reasons are put forward against this model. First, the authors doubt that the present management possesses the required skills to implement a major strategic change. Secondly, the lack of design and manufacturing of a CNC unit is looked upon as a major weakness for a merged UK industry. Thirdly, insufficient scale of output is seen as the overriding limitation to a UK merger solution for the machine tool industry.

Let us look at the second and third arguments. As regards the CNC unit, Sciberras and Payne argue that inhouse design and development of the CNC unit is important in the innovative process of many firms. As was argued in section 3.4, this was clearly the case for the two German firms following the focus strategy. Inhouse design and development of the CNC unit enables the firms to add functions to their lathes or systems that otherwise would not have been possible. Whilst there are clearly substantial benefits from having an inhouse CNC design, I need to point out that some own software development can take place on the basis of a bought standard CNC unit. For example, the leading UK firm, which equips its CNC lathes with robots of various types, buys in a standard CNC unit but makes its own software for connecting the robot to the CNC lathe. Nevertheless, substantial advantages can be gained from inhouse design, and any government policy in the UK should consider this issue very carefully.

According to Sciberras and Payne, the main problem is the small production volume that a merged UK firm could achieve. They argue:

> It may be more likely for a merged UK general purpose machine tool group to consider development and introduction of its own numerical controller. Volumes, however, will still be too small to enable cost effective competition compared to the large integrated US and Japanese firms. [Sciberras and Payne, 1985, p. 75]

In my opinion, the authors make comparisons with the wrong types of firms and this leads me to my critique of the perceived difficulties associated with a too small volume of output. It should, however, be noted that, according to Sciberras and Payne, their perspective is only a reflection of the views of the managers in the machine tool firms they interviewed. Hence, my critique applies equally to these managers.

Clearly, if one identifies the overall cost leadership strategic group as the one associated with production for the 'general purpose segment', then the problems of scale would be very considerable indeed. However, if instead one identifies the now-emerging new stategic group, which consists of the focus and the 'old' differentiation group, then the problem of scale becomes less important. This applies to both the CNC unit and CNC lathes proper.

In the case of the CNC unit, the leading firms in this new strategic group have their own CNC units, or at least have made the software themselves. These CNC units are different from those used by the firms following the overall cost leadership strategy or the low-performance strategy. Special functions are often included and the software is sometimes developed to suit particular groups of customers, e.g. firms producing oil-drilling equipment. The maximum *realized* scale economies for these more advanced CNC units come at an annual production level of approximately 1,000 units – this is the level of production of the leading German CNC lathe producer. The other two German firms, described in section 3.4, produce 500–600 units annually. Most interestingly, however, the Swedish firm claims that it produces its CNC units inhouse at a cost that is only half that asked by three leading (non-Japanese) specialized CNC producers. The Swedish firm had intended to design the main software itself and let one of these firms produce the CNC units, as in the case of two German firms. However, even at the volume of around 200 units per year, it proved to be cost efficient for the Swedish firm to continue to produce the CNC unit itself. Naturally, this does not mean that the unit cost would not decline substantially if its scale of output increased by, say, 300 per cent, but it shows that the question of scale economies is probably much less acute if a firm aims to join the strategic group of focus/differentiation rather than the overall cost leadership strategic group.

As for the production of the CNC lathes proper, Sciberras and Payne make the point that the three largest UK-owned CNC lathe producers produce only 35 CNC lathes per month. In 1984, the largest firm produced 190 units. This is of course not sufficient to achieve a competitive cost position. However, the comparison should be made not with the volumes of the largest Japanese firms of several thousand units but with those of the leading German producers of 500–600 units per year. Clearly, the largest UK firm could opt for the new strategic group of focus/differentiation, and this is even more the case if there were a merger of the largest firms. One can then conclude that a 'nationalistic' strategy of inducing the leading CNC lathe producers in both Sweden and the UK to join the new strategic group of focus/differentiation is not unrealistic. The firms should thus aim to meet the competition not from the firms within the overall cost leadership strategic group (mainly Japanese firms) but from those in the new strategic group of focus/differentiation. In this context it is also worth pointing out that, in terms of value, the German share of the UK market for CNC lathes was nearly as high as the Japanese in 1984 (23.5 and 28.4 per cent respectively). Furthermore, given the social benefits of a well-functioning machine tool industry, intervention to ensure a strategic reorientation of the firms would seem justifiable. Such an intervention presupposes, however, that the first problem posed by Sciberras and Payne – namely, poor management capabilities – is solved.

A strategic reorientation would have two main features. First, in order to become cost effective in the production of standard CNC lathes, volumes would need to be increased. Secondly, experience in system design needs to be generated at a pace that is as fast as or faster than their competitors'. As was argued in section 3.4, systems sales will, in all probability, develop into a volume market; in terms of the product cycle terminology, the system industry is now approaching the growth phase. As the domestic market is rather limited in Sweden and the UK, the firms will need to accumulate this experience mainly in foreign markets. The Swedish firm is already at this point.

As was argued above, a market expansion abroad is associated with higher costs and greater risks than expansion on the domestic market. These firms will need to be given the means to take the extra costs and risks associated with an expansion of system sales abroad. They will thus need to be given the resources to 'walk on two legs', i.e. with an expansion of standard CNC lathe production *and* of system sales. The alternative to implementing such a strategic reorientation will be a continued reliance on strategies that involve 'running away' from price competition with a concomitant poor development of production and employment. There is also, as was argued above, a very considerable risk that the present strategies of the Swedish and the UK firms will never be associated with reaching a break-even point financially, since the larger firms are now catching up quickly in terms of system sales and experience in system design. There is always, of course, the possibility that these smaller firms can leap to a different product, e.g. systems of larger size and integration. However, there are very considerable risks that the growing size of firms in the machine tool industry, in terms of both financial and technological capabilities, will mean that smaller firms, like the Swedish one, will not be able to survive even as a product developer in the future.

In terms of the content of such a government intervention, we can clearly learn a lot from the experience of the NICs. As in the case of these countries, a product-, firm- and function-specific policy ought to be implemented. The reasons for advocating such a policy in small industrialized countries are precisely the same – namely, that the social cost of transition to a viable industry is likely to be a lot lower than if general policies are pursued. The central feature of the policy would be the supply of enough risk capital to permit the firms to alter their strategic orientation. As in the case of the NICs, such a policy presupposes the existence of a competent and non-corrupt bureaucracy that can, together with industry and other actors, e.g. researchers, work out and agree upon the new strategic orientation.

9.5 Conclusions

There is ample evidence in the literature that, under some conditions, state intervention in the growth of industries may be justified on social grounds. In particular, imperfections in the capital market and the existence of external economies have been pointed to. The agreement on this issue is fairly widespread.

There is, however, less agreement in the literature on the question of the degree of uniformity/selectivity/custom design that these policies should have, and on the instruments that should be applied. One influential school of thought advocates the principle of uniformity of incentives (Balassa, 1975; Corden, 1980), whilst others, with more empirical knowledge, point to a need for selective and *custom-designed* intervention (Westphal, 1981; Katz, 1983b).

The question of which instrument to use is partly linked to the question of uniformity/selectivity. Balassa (1975) discards the theoretical justification for preferring subsidies to tariffs on account of the problems in the real world of budget constraints and an assumed relative unimportance of the welfare losses from price distortions. Corden (1980) argues for the existence of a hierarchy of policies, in apparent contradiction to his assertion of the superior qualities of uniform intervention. Baldwin (1969) argues, in a similar fashion to Corden, that the choice of instrument has to be determined by the specific problems of the particular industry.

Empirically, the three NIC countries studied have employed very different mixes of policy instruments to foster their machine tool industries. The difference is also great as regards their present policies vis-à-vis the CNC lathe industry. The performance of the industries varies substantially between the countries, as do the costs associated with fostering them. Although lacking data somewhat, I concluded that Taiwan has the highest benefit to cost ratio whilst Argentina has the lowest. Korea is in between. Strictly in terms of CNC lathe production, I would suggest that the ranking is still the same – Korea faring worse than Taiwan on account of import restrictions.

Coming back to the question of how an intervention should be designed, I agree with the notion that the type of intervention has to be product-specific. This is fairly unproblematic and obvious. My main criticism of the argument for *neutral* protection is that it may well be associated with a higher social cost of transition than a firm-specific policy. It is only when one is justified in assuming a homogeneity among firms (i.e. that one can analyse on the basis of a *representative* firm) that a *neutral* protection would be chosen in order to minimize the social cost of transition. Although in some industries such an assumption may be justified, it certainly is not in the CNC lathe industry or in other industries where product differentiation is an important feature of the industrial structure. In contrast, I suggested that a *strategy model* of firm behaviour, elaborated on in Chapter 3, would be closer to industrial reality. To the extent that this is true, not only product-, but also firm-specific policies would be appropriate. I further suggested that function-specific policies may be required. A blanket policy would be very limiting in terms of the type of strategy fostered. It might also be of little use because of discontinuities in firms' development.

The application of firm- and function-specific policies is well in line with Corden's (1980) proposition of a hierarchy of policies – given my assumption of a *non*-representative firm. It is also in line with Baldwin's (1969) claim that the policy instrument needs to be adapted to the specific problems of each industry.

Looking at the actual policies pursued by the three NIC governments in the area of CNC lathes, I concluded that both the Korean and Taiwanese governments' policies follow the conclusions drawn from a strategy model of firm behaviour. Product-, firm- and function-specific policies are implemented. In the case of Korea, however, severe trade restrictions are also included. This tool, I argued, is inefficient, costly and of marginal use to most firms. In the case of Argentina, the government continues with its prior policy of using trade restrictions as its main tool of intervention. Although this tool is, in this case, firm-specific as there is only one firm producing CNC lathes, the policy is for all practical purposes of a general nature. The prime characteristic of the policy, however, is that it was not selected on the basis of the needs of the firm; indeed, it had the totally opposite effect to creating an internationally competitive industry. In particular, it continues to be an important element behind the firm's orientation to the domestic market with all the negative effects on the benefit to cost ratio that such an orientation has been associated with in the past. Thus, the country with the highest benefit to cost ratio, Taiwan, has followed the policy prescriptions of the strategy model most closely. Korea, with a policy mix that partly departed from these prescriptions in that a general trade policy was part of the parcel, has a somewhat lower benefit to cost ratio. Argentina departs completely from these prescriptions and performs very badly.

In terms of the future, I would suggest that Taiwan and Korea continue with their selective policies and restrict the number of firms that can receive government subsidies to only two. I further suggest that these firms should be made to collaborate in the areas where economies of scale are important. The objective of government policy should be to help the firm(s) become a fully fledged member of the overall cost leadership strategic group. In the Argentinian case, the overall context for the firm has to be stabilized before an offensive industrial policy can be formulated.

Finally, in the case of the European industry, I would suggest that two firms, one Italian and probably one German, are induced to enter the overall cost leadership strategy so that Europe does not lose more market shares. I would also suggest that in the UK and Sweden, and possibly France, the state induces firms to reorient their strategy towards the new merging strategic group of focus/differentiation. I suggest that a *nationalistic* policy is viable along this line. As in the case of the NICs, the government policy should be product-, firm- and function-specific in nature.

Notes

1 Other types of external economies also operate, e.g. the *infant economy* argument. See, for example, Kreuger (1981b) and Corden (1980).
2 A protectionist system does not necessarily mean only tariff protection. As Corden himself puts it: 'Effective protection may be interpreted more broadly than just being concerned with tariffs or with the overall protection for import competing industries. It can also refer to protection for exporting activities, and I can allow for export subsidies, export taxes, and indeed for all sorts of devices, taxes and subsidies' (1980, p. 71).
3 I am most grateful to Dr Kim Seung Jin for clarifying the credit system in Korea.
4 The market price in the USA of the CNC lathe is roughly the same as the market price in Sweden of the Japanese lathe that was used as a model for the Korean design efforts. Performance-wise, the CNC lathes are similar.
5 These figures may understate the true subsidy. An element of cross-subsidy may also exist in conventional machine tools, which constitute the bulk of the output of the firm. Although I have no data to calculate the size of such a subsidy, I would not expect it to be high as the conventional machine tools produced are simple and therefore the firm can quickly learn how to produce them at an international cost. Furthermore, there are many producers of these machine tools in Korea, making it very unlikely that competition is not fierce.
6 I assume that the total lathe production amounted to roughly 10–15 million USD in Argentina in these years. This would be equal to approximately 25 per cent of the total machine tool production in Argentina, which was the ratio of lathes to all machine tools produced (in value) about 1970.
7 This section is based on Barnevik (1985) and Sigurdson (1985).

10

Summary

This book provides an analysis of the world computer numerically controlled (CNC) lathe industry. This is an industry that has undergone far-reaching structural changes since 1975. The book analyses the process of structural change, the adjustment of firms within the industry, and the role of governments in assisting firms to adjust. The countries investigated in the book are both the Newly Industrializing Countries (NICs) Argentina, Korea and Taiwan and the developed countries in Europe as well as Japan.

The broader aim of the book was to contribute to two debates that formed the intellectual origin of the research question – the debate on how the *electronic revolution* affects the structure of the international industry, as well as to the debate on how governments should design their policies in order to improve their industry's position in the international division of labour.

Four areas within the general research question were examined:

- the international diffusion of CNC lathes (Chapter 2);
- the international CNC lathe producing industry (Chapters 3 and 4);
- the experiences of the three NICs (Chapters 5–8);
- the role that government policies have had in the adjustment processes of the NIC-based firms and the lessons one could draw from the case studies with regard to the design of policy interventions; the future strategic orientation of the European industry was also discussed and how government policy should be designed in these countries (Chapter 9).

10.1 The diffusion of CNC lathes

The process of turning is undergoing fundamental changes in the technology used. CNC lathes are replacing conventional lathes and, in 1984, CNC lathes accounted for over 80 per cent of the investment in lathes in some developed countries. One consequence of this substitution process has been a rapid decline in the market for the simpler types of lathes that the NICs have specialized in producing and exporting to the developed country markets. This change is in itself sufficiently alarming for the lathe producers in the NICs to contemplate adjusting and starting to produce CNC lathes too. In addition, there is a widespread diffusion of CNC lathes in the three NICs in question, which further reduces the market for conventional lathes. The lathe producers in the three NICs under study are aware of the threat that this process poses to them and some of them are attempting to adjust by moving into the production of CNC lathes.

10.2 The international CNC lathe producing industry

The industry that these NIC firms are attempting to enter underwent a drastic structural change in the second half of the 1970s and the early 1980s. The Japanese firms became, very suddenly, the world leaders and captured 50 per cent of the non-Japanese market. The leading Japanese firms are today producing ten times as many CNC lathes per annum as the globally leading firms did in the mid-1970s. A tentative analysis of the reasons for this Japanese success was given in Chapter 3. I suggested that the risks involved in the strategic reorientation by the Japanese firms in the mid-1970s were lower than they would have been for the European and US firms. In particular, I noted an early orientation towards a demand for less complex CNC lathes by the Japanese industry and lack of import competition as two risk-reducing factors. The European responses to this Japanese expansion vary very considerably. The odd firm has copied the Japanese strategy. Other firms adjusted by adapting their products and increasing the volume of output so as to become more price competitive. Yet others ran away from price competition altogether.

In the early 1980s, a new pattern within the industry had materialized. I identified three strategic groups within the industry and outlined the main characteristics of each group. The main barriers to entry into each group were also identified. I then discussed (Chapter 4) the barriers to entry into the dominant strategic group – namely, that consisting of the firms following the *overall cost leadership strategy*. I discussed the size and character of the barriers to entry in the areas of R&D, procurement of components, manufacturing, and marketing and after-sales service. I also attempted to specify the minimum efficient scale of production and found it approached 1,000 units per annum.

10.3 The NIC experience

I then turned to analyse the machine tool industries of Argentina, Korea and Taiwan and the progress they had made in the field of CNC lathes. Eight firms were studied in detail. Broadly speaking, these firms constitute a fourth strategic group within the CNC lathe industry. The leading firms within this group are one Korean and one Taiwanese firm. These two countries are also way ahead of other NICs in this field. I argue, however, that these eight firms are in a strategic position that is untenable in the long run, in particular because they cannot reap the full benefits of scale economies when they follow this particular strategy. A shift to another strategy is required, and I argued that the most suitable one would be the overall cost leadership strategy, i.e. the one that is followed mainly, but not exclusively, by the world leading Japanese firms.

Detailed chapters (6, 7 and 8) followed the broader introductory Chapter 5. In each of these chapters I analysed the context in which the firms operate. In particular, I analysed the size and character of the engineering industry, the government policies and the local demand for CNC lathes. An account of the firms – their history and strategic position – was also given. Finally, I discussed present government policies and their appropriateness with respect both to the needs of industry and to social efficiency. The three countries are similar only with respect to the size of the domestic market. In all of them, the number of units sold in the domestic market annually is very small in relation to the minimum efficient scale of production. They differ widely with respect to government policies. The Argentinian governmental policies have been characterized by high tariff barriers and a very fluctuating real exchange rate resulting in an inward orientation not only of the firm analysed but also of the whole industry. A high export subsidy failed to induce the firms to export. The Taiwanese case shows little state intervention until very recently. A generally more stable environment than in Argentina has, however, had important effects on the strategic choice of the machine tool firms. Recently, the state has begun to use the credit instrument as a means to influence the firms' strategic choices. The Korean case is again very different. The state has intervened heavily in the machine tool industry, although the instruments have been somewhat different from the Argentinian case. Strict import controls have been accompanied by a very generous credit policy. The state, furthermore, emphasized the necessity to export in its discussions with the firms. I attempted to trace the links between the different contexts, mainly shaped by these government policies, and the strategic choices made by the firms. The historical strategic choices can then account for the present capabilities of the firms.

10.4 Government policies

In the literature, there is a widespread agreement that state intervention in the growth of industries may be justified on social grounds. Imperfections in the capital market and the existence of external economies have, in particular, been singled out. Less agreement exists however on the issue of *how* the state should intervene. The main battle lies between those advocating general policies and those advocating selective policies. The problem of which instrument(s) to use in the intervention is partly linked to the question of uniform versus selective policies. One school advocates blanket policy prescriptions whilst another argues for the existence of a hierarchy of policies.

In my view it is obvious that the intervention needs to be *product* specific. My main criticism of the notion of uniform protection between firms is that it may well be associated with a higher social cost of transition than a firm-specific policy. It is only when one is justified in assuming homogeneity among firms that a uniform or neutral protection would be chosen in order to minimize the social cost of transition. I have demonstrated in detail that this assumption is wrong in the case of the CNC lathe industry; I instead used a strategy model of firm behaviour to analyse the industry. On the basis of the strategy model, I showed that not only firm-specific but also function-specific policies would be more appropriate. Furthermore, I suggested that a blanket policy would be very limiting in terms of the type of strategies fostered. It may also be of little use on account of discontinuities in firms' development.

Looking at the actual policies pursued by the governments of the three NICs, the country with the highest benefit to cost ratio for the machine tool industry – Taiwan – has followed the policy prescriptions of the strategy model most closely. Korea has a somewhat lower benefit to cost ratio on account of the implementation of general trade restrictions. Argentina departed fully from these prescriptions and performed very badly.

In terms of the future, the objectives of the policies of the governments of Korea and Taiwan should be to help at most two of their leading firms to become full-fledged members of the strategic group of overall cost leadership. The governments should continue using selective policies and restrict the number of firms that receive government subsidies to at most two. In the Argentinian case, the overall context of the firm has to be stabilized before an offensive industrial policy can be formulated.

In the case of the European industry, two firms, probably one Italian and one German, should be induced to pursue the overall cost leadership strategy so that the European industry can keep its market share. I also suggest that the states in Sweden, the UK and possibly France induce firms to re-orient their strategies towards the new, emerging strategic group of focus/differentiation where firms combine system design capability with medium volume production of stand alone, standard, CNC lathes. As in the case of the NICs, the government policy should be product, firm and function specific in nature.

References

Amadeo, E. P., Fernandez, R. F. and Morales, F. (1982), *Argentine's Machine Tool Sector*, IDRC-MR 35e (Ottawa: IDRC).
American Machinist (1979), December (New York: McGraw-Hill).
American Machinist (1983), February and November (New York: McGraw-Hill).
Amsden, A. (1977), 'The division of labour is limited by the type of market; the case of Taiwanese machine tool industry,' *World Development*, vol. 5(3), March, pp. 217–33.
Ansoff, H. I. (1977), 'Towards a strategic theory of the firm,' in H. I. Ansoff (ed.), *Business Strategy* (Harmondsworth, Middx: Penguin).
Arnold, E. (1983), 'Competition and technological change in the UK television industry,' a thesis submitted to the University of Sussex in partial fulfilment of requirements for the degree of Doctor of Philosophy.
Arocena, J. S. de (1981), 'Notas sobre la Informatica en Venezuela' (mimeo, Caracas: Cendes, UCV).
Arrow, K. (1962), 'The economic implications of learning by doing,' *Review of Economic Studies*, June.
Association of Argentinian Engineering Industries (1980), Interviews.
Balassa, B. (1985), 'Reforming the system of incentives in developing countries,' *World Development*, vol. 3(6), June, pp. 365–82.
Balassa, B. (1981), *The Newly Industrializing Countries in the World Economy* (New York: Pergamon Press).
Baldwin, R. (1969), 'The case against infant industry tariff protection,' *Journal of Political Economy*, vol. 77, pp. 295–305.
Banco Central de la Republica Argentina (no date), *Estimaciones trimestrales sobre oferta y demanda global*.
Bank of Korea, The (1984), *Economic Statistics Yearbook 1984* (Seoul).
Barnevik, P. (1985), 'Who are the innovators? A multinational viewpoint. The case of industrial robots,' Paper delivered to the *Financial Times* and the Institute for Research and Information on Multinational's Conference 'Multinationals: Innovators in High Technology', Munich, 24–25, April 1985.
Barron, I. and Curnow, R. E. (1978), *The future of information technology. Forecasting the effects of information technology* (London: Francis Pinter Publishers Ltd).
Bell, M., Ross-Larsen, B. and Westphal, L. (1983), 'The cost and benefit of infant industries. A summary of firm-level research (mimeo, Science Policy Research Unit, University of Sussex).
Bendix, P. J., Kim, C., Körner, M., Kloos, U., Schneider, K. and Wolff, P. (1978), *Development and Perspectives of the Korean Machinery Industry – with Special Reference to Machine Tool, Electrical Machinery and Plant Equipment Manufacturing* (Berlin: German Development Institute).
Bhagwati, J. and Srinasasan, T. N. (1978), 'Trade policy and development,' in R. Cornbusch and J. Frenkel (eds), *International Economic Policy: Theory and Evidence* (Baltimore, Md: Johns Hopkins University Press).

REFERENCES

Bhattacharay, S. K. (1976), 'Penetration and utilization of NC/CNC machine tools in British industry' (mimeo, Birmingham: Department of Engineering Production, University of Birmingham, September).

Boon, G. K. (1984), *Flexible Automation: A comparison of Dutch and Swedish firms particularly as to CNC machine penetrations*, Working paper TSF 84–2 (Noordwijk, Holland).

Boston Consulting Group, The (1985), *Strategic Study of the Machine Tool Industry, Summary Report* (Dusseldorf).

Calvar, J., Sallaberry, N. and Monteverde, E. (1980), 'Enfoque de la Participacion del Sector Publico a la Demanda de Bienes y Servicios Disponibles a Precios Constantes (Period 1970/78),' *Series de trabajos metodologicos y sectorales*, no. 7, February (Buenos Aires: Banco Central de la Republica Argentina).

CAFMHA (Camara Argentina de la Maquina Herramienta) (1979), *Statistical Figures of Argentina Machine Tools* (Buenos Aires).

Canitrot, A. (1981), 'Teoria y practica del liberalismo. Politica antiinflacionaria y apertura economico en la Agentina, 1976–1981,' *Desarollo Economico*, vol. 21(82), pp. 130–89.

Castano, A., Katz, J. and Navajas, F. (1981), *Etapas historicas y conductas tecnologicas en una planta Argentina de maquinas herramientas*, Monografia de Trabajo No. 38, January (Buenos Aires: BID/CEPAL/CIID/PNUD Programa de Investigaciones Sobre Desarrollo Cientifico y Tecnologico en America Latina).

Caves, R. E. (1980), 'Industrial organization, corporate strategy and structure,' *Journal of Economic Literature*, vol. XVIII (March), pp. 64–92.

Caves, R. E. and Porter, M. (1977), 'From entry barriers to mobility barriers: conjectural decisions and contrived deterrence to new competition,' *Quarterly Journal of Economics*, vol. XLI(2), May, pp. 241–61.

Center on Transnational Economy (1984), 'The diffusion of electronics technology in developing countries' capital goods sector: The Argentinian case (mimeo, Buenos Aires).

CEPD (Council for Economic Planning and Development) (1975), *Industry of Free China*, no. 6.

CEPD (1980), *Industry of Free China*, no. 12.

CEPD (1981), *Four-Year Economic Development Plan for Taiwan, Republic of China (1982–1985)* (Taipei).

CEPD (1982), *Taiwan Statistical Data Book*.

CEPD (1983), *Industry of Free China*, no. 5.

CEPD (1985), *Industry of Free China*, no. 6.

Chamberlain, E. H. (1960), *The Theory of Monopolistic Competition. A Reorientation of the Theory of Value*, 7th edn (Cambridge, Mass.: Harvard University Press).

Chou, K. K. (1982), 'A comparison of the machine tool industries in the ROC and the ROK,' *Taiwan Industry: Machine Tools*, no. 8, pp. 44–51.

Chou, K. K. (1983), Interview.

Chudnovsky, D., Nagao, M. and Jacobsson, S. (1983), *Capital Goods Production in the Third World. An Economic Study of Technology Acquisition* (London: Francis Pinter).

Cooper, C. (1976), 'Policy intervention for technological innovation in less-developed countries' (mimeo, Brighton: University of Sussex).
Cooper, C. and Hoffman, K. (1978), 'Transactions in technology and implications for developing countries' (mimeo, Brighton: Science Policy Research Unit and Institute of Development Studies, University of Sussex).
Corden, W. H. (1974), *Trade Policy and Economic Welfare* (Oxford: Oxford University Press).
Corden, W. H. (1980), 'Trade policies,' in J. Cody, H. Hughes and D. Wall. *Policies for Industrial Progress in Developing Countries* (Oxford: Oxford University Press).
Corden, W. H. (1981), 'Comments,' in W. Hong and L. Krause, *Trade and Growth of the Advanced Developing Countries in the Pacific Basin* (Seoul: Korea Development Institute).
Daewoo Heavy Industries Ltd (no date), Information pamphlet (Seoul).
DEK (1981), *Datateknik i verkstadsindustrin. Datorstödd konstruktion och tillverkningsteknik*, SOU 1981:10 (Stockholm: Liber Förlag).
DGFM (Direccion General de Fabricaciones Militares) (1972), *Estructura de la Industria de Maquinas Herramientas* (Buenos Aires).
Digest of Japanese Industry and Technology (1982), 'A survey of machine tool installations in Japan,' no. 176.
Dytz, E. (1982), 'Controle de processos e automatizacao industrial,' in *Maquinas & Ferramentas*, May.
Economic Planning Board, The Republic of Korea (various), *Report on Mining and Manufacturing Survey* (Seoul).
Edquist, C. and Jacobsson, S. (1982), *Technical Change and Patterns of Specialization in the Capital Goods Industries of India and the Republic of Korea – a project description* (Lund: Research Policy Institute, University of Lund).
Edquist, C. and Jacobsson, S. (1985a), *Trends in the Diffusion of Electronics Technology in the Capital Goods Sector*, Discussion Paper No. 161 (Lund: Research Policy Institute).
Edquist, C. and Jacobsson, S. (1985b), 'The production of hydraulic excavators and machining centres in India and the Republic of Korea,' Paper delivered to the ICSSR/SAREC Conference on Processes of Industrialization and Technological Alternatives, New Delhi, 1985.
Edquist, C. and Jacobsson, S. (1986), 'The impact on developing countries of automation in the capital goods industry' (mimeo, Lund: Research Policy Institute).
Elsässer, B. (1983a), 'Numeriskt styrda verktygsmaskiner och produktivitetsutveckling i maskinindustrin 1969–1980' (mimeo, Linköping: Linköpings Universitet).
Elsässer, B. (1983b), Interview.
Elsässer, B. and Lindvall, J. (1984), *Numeriskt styrda verktygsmaskiner. Effekter på produktivitet, arbetsvolym, yrkesstruktur och kvalifikationskrav* (Linköping: Ekonomiska Institutionen, Universitetet i Linköping).
Engineer, The (1975), 15 May.
Engineer, The (1979), 22 February.
Ergas, H. (1984), *Why do some countries innovate more than others?* Centre

for European Policy Studies, CEPS Papers No. 5 (Brussels).
Eurostat (various years), *NIMEXE Analytical Tables of Foreign Trade* (Luxemburg).
Export–Import Bank of China, The (1981), *Annual Report* (Taiwan).
Financial Times (1982), 19 January.
Financial Times (1983), 7 April and 27 September.
Freeman, C. (1977), 'Technical change and unemployment,' Paper presented to conference on 'Science, Technology and Public Policy – an international perspective,' 1–2 December (Kensington: University of New South Wales).
Galbraith, C. and Schendel, D. (1983), 'An empirical analysis of strategy types,' *Strategic Management Journal*, vol. 4, pp. 153–73.
Gebhardt, A. and Hatzold, O. (1974), 'Numerically controlled machine tools,' in L. Nabseth and G. F. Ray (eds), *The Diffusion of New Industrial Processes* (London: Cambridge University Press).
Gildemeister Aktiengesellschaft (1979), *Annual Report* (Bielefeld).
Gilmore, F. F. and Brandenburg, R. G. (1977), 'Anatomy of corporate planning,' in H. I. Ansoff, *Business Strategy* (Harmondsworth, Middx: Penguin).
Gould, B. (1981), 'Changing perspectives on size, scale and returns: An interpretive survey,' *Journal of Economic Literature*, vol. XIX (March), pp. 5–33.
Göransson, B. (1984), *Enhancing National Technological Capability. The Case of Telecommunications in Brazil*, Technology and Development Discussion Paper No. 158 (Lund: Research Policy Institute, University of Lund).
Hirsch, S. (1967), *Location of Industry and International Competitiveness* (Oxford: Clarendon Press).
Hoffman, K. and Rush, H. (1980), 'Microelectronics, industry and the Third World,' *Futures*, August.
Hufbauer, G. C. (1966), *Synthetic materials and the theory of international trade* (London: Duckworth).
IDIC (Industrial Development and Investment Center) (1982a), *Applicable Scope of the Strategic Industry* (Taipei).
IDIC (1982b), *Taxes in Taiwan* (Taipei).
INTI (Instituto Nacional de Technologia Industrial) (1981), *Estudio sobre el desarollo de la industria electronica Argentina* (Munich: Ministerio de Investigacion y Tecnologia de la Republica Federal e Alemania).
Jacobsson, S. (1982a), 'Electronics and the technology gap – the case of numerically controlled machine tools,' in R. Kaplinsky, 'Comparative advantage in an automating world,' *IDS Bulletin*, vol. 13(2), March.
Jacobsson, S. (1982b), *Technical Change and Technology Policy. The Case of Numerically Controlled Lathes in Argentina*, Working Paper No. 44 (Buenos Aires: ECLA/IDB/IDRC/UNDP Research Programme on Scientific and Technological Development in Latin America).
Jacobsson, S. and Sigurdson, J. (eds) (1983), *Technological Trends and Challenges in Electronics* (Lund: Research Policy Institute, University of Lund).
Japan Economic Journal (1981) 9 June and 23 June.
Japan Tariff Association (1983, 1984, 1985), *Japan Exports and Imports*.

Jenkins, C. and Sherman, B. (1979), *The Collapse of Work* (London: Methuen Eyre).
JMTBA (Japan Machine Tool Builders' Association) (1980), *Machine Tool Industry, Japan* (Tokyo).
Jones, D. (1983), 'Machine tools: Technical change and the Japanese challenge' (mimeo, Brighton: Science Policy Research Unit, University of Sussex).
Kaplinsky, R. (ed.) (1982a), 'Comparative advantage in an automating world,' *IDS Bulletin*, vol. 13(2).
Kaplinsky, R. (1982b), *The Impact of Microelectronics and the International Division of Labour: An Illustrative Case Study of Computer Aided Design* (Vienna: Industrial Studies Division, UNIDO).
Katz, J. (1976), *Importacion de Tecnologia, Aprendizaje Local e Industrializacion Dependiente* (Mexico: Fondo de Cultura Economica).
Katz, J. (1983a), 'Domestic technological innovation and dynamic comparative advantages: Further reflections on a comparative case-study programme' (mimeo, Buenos Aires: ECLA).
Katz, J. (1983b), 'Technological change in the Latin American metalworking industries. Results of a programme of case studies,' *CEPAL Review*, April (Santiago, Chile).
Kim Lin Su (1982), *Technological Innovations in Korea's Capital Goods Industry: a micro analysis* (Geneva: Technology and Employment Programme, ILO).
Kim Seung Jin (1983a), *Evaluation of and Reform Proposals for Promotion Policies in the Korean Machinery Industry*, Working Paper 83–06 (Seoul: Korea Development Institute).
Kim Seung Jin (1983b), 'Patterns of production and trade in machine tool products in the Pacific Basin and implications for regional cooperation' (mimeo, Seoul: Korea Development Institute).
KMTMA (Korea Machine Tool Manufacturers' Association) (1983), *Machine Tool*, 4.
KMTMA (1984), *Machine Tool*, 5.
KMTMA (1985a), *Machine Tool*, 9.
KMTMA (1985b), *Machine Tool*, 11.
Korea Annual (1982), Yonhap News Agency (Seoul).
Kreuger, A. (1981a), 'Comments,' in W. Hong and L. Krause (eds), *Trade and Growth of the Advanced Developing Countries in the Pacific Basin* (Seoul: Korea Development Institute).
Kreuger, A. (1981b), 'Export-led industrial growth reconsidered,' in W. Hong and L. Krause (eds), *Trade and Growth of the Advanced Developing Countries in the Pacific Basin* (Seoul: Korea Development Institute).
Lahera, E. and Nochteff, H. (1983), 'La microelecronica y el desarollo Latinamericano,' *Revista de la CEPAL* (Santiago, Chile).
Lall, S. (1978), 'Developing countries as exporters of technology' (mimeo, Oxford: Institute of Economics and Statistics, University of Oxford).
Landes, D. (1969), *The Unbound Prometheus* (Cambridge: Cambridge University Press).
Lee, B. J. (1983), Interview.

Liang Kho-Shu, Liang Ching-Ing Hou (1981), 'Trade strategy and exchange rate policy in Taiwan,' in W. Hong and L. Krause (eds), *Trade and Growth of the Advanced Developing Countries in the Pacific Basin* (Seoul: Korea Development Institute).
Little, I., Scitovsky, T. and Scott, M. (1970), *Industry and Trade in Some Developing Countries. A Comparative Study* (London: Oxford University Press).
Ljung, T. (1980), *Automationsbenägenheten i svensk verkstadsindustri – spridning och användning av flexibel automation* (Lund: Lund Institute of Technology).
McGee, J. S. (1974), 'Efficiency and economies of size,' in H. J. Goldschmid, H. M. Mann and J. F. Weston, *Industrial Concentration: the new learning* (Boston: Little, Brown).
McLean, J. M. and Rush, H. J. (1978), *The impact of microelectronics on the UK. A suggested classification and illustrative case studies*, SPRU Occasional Papers No. 7 (Brighton: Science Policy Research Unit, University of Sussex).
Marx, K. (1974), *Capital*, vol. One (London: Lawrence and Wishart).
MCI (Ministry of Commerce and Industry) (1981), *1981 Basic Plan for the Advancement of the Machinery Industry* (Seoul).
MCI (1983), Interview (Seoul).
Metal Industrial Development Centre (1982), *Machine Tools*, April (Taipei).
Metalworking, Engineering and Marketing (1979), September.
Metalworking, Engineering and Marketing (1980), March.
Metalworking, Engineering and Marketing (1982), May.
Metalworking, Engineering and Marketing (1983a), January.
Metalworking, Engineering and Marketing (1983b), July.
Metalworking, Engineering and Marketing (1985), September.
Metalworking Production (1979), April.
Ministerio de Economia (1979), 'Panorama del sector metalmecanico Argentino' (mimeo, Buenos Aires).
MTTA (Machine Tool Trades Association) (1983 and 1985), *Machine Tool Statistics* (London).
Nam, C. H. (1981), 'Trade, industrial policies, and the structure of protection in Korea,' in W. Hong and L. Krause (eds), *Trade and Growth of the Advanced Developing Countries in the Pacific Basin* (Seoul: Korea Development Institute).
Needham, D. (1978), *The Economics of Industrial Structure and Performance* (London: Holt, Rinehart and Winston).
NMTBA (National Machine Tool Builders' Association) (1981/82, 1983/84, 1984/85, 1985/86), *Economic Handbook of the Machine Tool Industry* (Virginia).
Noble, D. F. (1978), 'Social choice in machine design: The case of automatically controlled machine tools and a challenge for labour,' *Politics and Society*, vol. 8 (3–4).
OECD (1983), 'Competitive du system productif et specialisation internationale' (mimeo, Paris).
Office of Customs Administration (various), *Statistical Yearbook of Foreign Trade* (Seoul).

Oriental Economist, The (various years), *Japan Company Handbook* (Tokyo).
Patrick, H. (1981), 'Comments to Liang and Liang,' in W. Hong and L. B. Krause (1981), *Trade and Growth of the Advanced Developing Countries in the Pacific Basin* (Seoul: Korea Development Institute).
Planning Research & Systems Limited (1979), *A market study of numerically controlled machine tools in the United Kingdom, France, Italy and Germany* (London: John Martin Publishing Limited).
Porter, M. E. (1980), *Competitive Strategy. Techniques for Analyzing Industries and Competitors* (New York: The Free Press).
Rada, J. (1980), *The impact of microelectronics. A tentative appraisal of information technology*. (Geneva: International Labour Office).
Rattner, H., Machline, C. and Udry, O. (1981), 'Production and diffusion of numerically controlled machine tools in Brazil' (mimeo, Sao Paolo: EAESP/FGV).
Rattner, H. (1984), *La difusion de maquinas-herramienta de control numerico en Brazil*, The Technology Scientific Foundation, Working Paper No. TSF 84–4 (Holland).
Rempp, H. (1982), 'Introduction of CNC machine tools and flexible manufacturing systems: Economic and social impact,' Paper presented at a conference on European Employment and Technological Change, Rome, 10–12 February.
Research Policy Institute (1983), *Technology and Development* (Lund: University of Lund).
Rosenberg, N. (1976), *Perspectives on Technology* (Cambridge: Cambridge University Press).
Samuelson, P. A. (1964), *Economics* (New York: McGraw-Hill).
Scherer, F. M. (1980), *Industrial market structure and economic performance* (Chicago: Rand McNally College Publishing Company).
Sciberras, E. and Payne, M. (1985), *The UK Machine Tool Industry: Recommendations for Industrial Policy* (London: The Technical Change Centre).
Scott, M. (1979), 'Foreign trade,' in W. Galenson (ed.), *Economic Growth and Structural Change in Taiwan. The Postwar Experience of the Republic of China* (Ithaca: Cornell University Press).
Senker, P., Swords-Isherwood, N., Brady, T. and Hugget, C. (1980), 'Maintenance skills in the engineering industry. The influence of technological change' (mimeo, Brighton: Science Research Policy Unit, University of Sussex).
Sigurdson, J. (1985), 'Developing a production and marketing capability for robots – the case of ASEA in Sweden' (mimeo, Lund: Research Policy Institute).
Smith, A. (1977), *The Wealth of Nations* (Harmondsworth, Middx: Pelican).
SMT Machine Company (various years), *Annual Reports* (Västerås).
Soete, L. G. (1981), 'A general test of technological gap trade theory' (mimeo, Brighton: Science Policy Research Unit, University of Sussex).
Sourrouille, J. V. and Lucangeli, J. (1980), 'Apuntes sobre la historia reciente de la industria Argentina. Los resultados del censo industrial de 1974' (mimeo, Buenos Aires: ECLA).

Statistical Yearbook (1984) of the Republic of China.
Statistiches Bundesamt (1981), *Aussenhandel*, Reihe 2, 'Specialhandel nach waren und Landeren 1980' (Wiesbaden).
Steen, H. (1976), *Rapporter från Sveriges Mekanförbund*, 10 August and 21 October.
Steen, H. (1977), *Rapporter från Sveriges Mekanförbund*, 8 September.
Stemmer, C. E. (1981), 'Estagio atual do comando numerico no Brasil,' in *Maquinas & Ferramentas*, May.
Syndicat des Constructeurs Français de Machine Outils (1976, 1980), *Statistiques Commerce Extérieur*.
Syndicat des Constructeurs Français de Machine Outils (1977, 1981), *Enquête Syndicale sur la commande numérique dans l'industrie de la Machine Outil*.
Tauile, J. R. (1981), 'Introductory notes to the political economy of information' (mimeo, Rio de Janeiro: Universidade Federal de Rio de Janeiro, New School for Social Research).
Today's Machine Tool Industry (1976), 'Present situation and future trends in NC industry in Japan' (Tokyo: News Digest Publishing, International Edition).
Today's Machine Tool Industry (1977a), 'Progressive development of numerically controlled machine tools in Japan' (Tokyo: News Digest Publishing, International Edition).
Today's Machine Tool Industry (1977b), 'The recent technical trends in NC lathes' (Tokyo: News Digest Publishing, International Edition).
Tsai, J. (1983), Interview.
UNCTAD (1982a), *Problems and Issues Concerning the Transfer, Application and Development of Technology in the Capital Goods and Industrial Machinery Sector. The Impact of Electronics Technology on the Capital Goods and Industrial Machinery Sector: Implications for Developing Countries*, TD/B/C.6/AC.7/3 (Geneva).
UNCTAD (1982b), Secreteriat Data Bank Compilations based on data from the Statistical Office of the United Nations, March (Geneva).
United Nations (1979/80), *Statistical Yearbook* (New York).
United Nations (1980a), *Yearbook of National Accounts Statistics* (New York).
United Nations (1980b), *Yearbook of Industrial Statistics*, vol. I (New York).
Utterback, J. M. (1979), 'The dynamics of product and process innovations in industry,' in C. T. Hill, and J. M. Utterback, *Technological Innovation for a Dynamic Economy* (Oxford: Pergamon Press).
Valeiras, J., Carqiulo, G., Vertiz, L., Trigueros de Godey, E., Dorrego, E. and Pelos, L. (1978), *Una politica nacional de maquinas-herramienta* (Buenos Aires: CAFMHA).
Watanabe, S. (1983), *Market Structure, Industrial Organization and Technological Development: the case of the Japanese electronic-based NC-machine tool industry*, Working Paper, WEP 2–22/WP111 (Geneva: ILO).
Westphal, L. (1978), 'Industrial incentives in the Republic of China (Taiwan)' (mimeo, Washington, DC: The World Bank).
Westphal, L. (1981), *Empirical Justification for Infant Industry Protection*, World Bank Staff Working Paper No. 445. (Washington, DC: The World Bank).

World Bank (1979), *Argentina. Structural Changes in the Industrial Sector*, Report No. 1977a-AR, 30 March (Washington, DC).

World Bank (1985), *World Development Report 1985* (Washington, DC).

Yuravlivker, D. E. (1980), 'Political shocks and the real exchange rate in a developing country – the case of Argentina,' Paper presented at the Joint Workshop in International Economics and Public Finance and Latin America Workshops, Buenos Aires, 11 June.

Index

aerospace (aircraft) 54, 55
after-sales service 95–7, 101 n.6
Argentina 1, 3, 4, 22–30 *passim*, 89, 102–7 *passim*, 111, 113–42
  CNC industry compared with Sweden 125–7
  engineering sector 113, 114 (table 6.1), 115 (table 6.2)
  government policy 113, 119–21, 193–219 *passim*
    export subsidies 214
    licensing 139–40
    machine tool, exports 203 (table 9.2)
    marketing 135, 141
    skills, shortage of 123–4, 135, 141
    tariff barriers 113, 115, 116, 119, 120, 137–41 *passim*, 202, 207, 219, 233
ASEA (Sweden) 45, 224–5
Asia(n) 91 (table 4.1), 102–3, 110, 131, 224
Australia 34
automation 166
automobile industry 54, 67, 77, 81

barriers to entry, *see* entry
batch production 8
batch size 14–15, 26
  reduction of 22
Brazil 30, 102–3, 130, 136, 139

China External Trade Development Council (CETDC) 167–8, 173
Cincinnati Milacron 80
circuitry, handwired 9
CNC (computer numerical control) lathes 7, 8, 47
  components 91–4, 99, 107
  costs *see also* depreciation 13, 14, 15, 19–30 *passim*, 55, 201, 205–8, 230–1
    capital (investment) 13, 15, 19, 21, 22–3, 98

CNC (computer numerical control) lathes – *contd.*
  costs *see also* depreciation – *contd.*
    labour 13, 98; saving of 20 (tables 2.9, 2.10), 21 (table 2.11), 23
    skills, reduction in 23–8
    stocks and work in progress 21 (table 2.11), 22–3
    transport 97
  custom-designed 71, 76, 77, 110, 209, 210
  distributors, local independent 96, 110
  imports into OECD 68 (table 3.18)
  imports into USA 79–80
  Japanese, increase in production 31–7 *passim*, 47–66
    European response 66–79
  lifetime of 89
  machine-tool building, links with 56
  maintenance 23, 96
  manufacturing 94–5
    in 5 NICs 103, 203, 204–5, *see also under separate countries*
  operations, shortage of 24–5
  price discrimination 91–2
  production, growth of 31–7
  programmers 23, 24
  programming time 15
  repairs 23, 96
  S-curve 43 (Fig 3.1), 45, 46, 47
  setters 23, 24
  spindle, multi 70
  spindle, single 70
  submarkets 66–7
  substitution for conventional lathes 29
  technology 8–12
  units
    fixed feature 9
    fully flexible 9
    microcomputer-based 57
  weight, relative, per unit 48 (table 3.7)
Comecon countries 36, 47
Computer Aided Design unit 156
Computerized Parts Changer (CPC) 72–6 *passim*

depreciation 19, 22, 26 (tables 2.12 and 2.13), 28 (tables 2.16, 2.17)
design 95
　engineers, *see* engineers
　skill requirements 90, 110, 171, 187
　unit, computer-aided 156
designers 76–7, 108, 109, 110, 135, 156, 161–3 *passim*, 171
　core 90, 108, 135
dynamic external economies 196–7
dynamic internal economies 195–6

EEC (European Economic Community) 36, 47, 58, 76
engineers
　design 70, 106 (table 5.4), 109, 131, 171, 186, 188, 206–7, 210, 214, 218
　electronic 90, 106 (table 5.4), 108, 109, 131, 218
　mechanical 218
　production 228
　service 159
engines 8
entry into industry, firms' 41, 109
　barrier to 41, 75, 83, 85 (table 3.23), 86, 88–101, 105, 109, 110, 154, 209, 210
Europe(an) 4, 31–7 *passim*, 47, 52, 53, 58–63, 66–85 *passim*, 89, 92, 97, 100, 221–32
　challenge 67
　CNC industry (options) 221–32
　Japanese competition, response to 66–79

Fanuc Fujitsu (Fanuc) 9, 56–8, 85, 86 n.2, 91–2, 93, 112 n.2, 162, 187
Fiat 67
FMM (Flexible Manufacturing Module) 12, 67–72 *passim*, 75, 84 (table 3.22), 95, 209, 222, 227
FMS (Flexible Manufacturing Systems) 95, 186
France 11, 13, 17, 58–61 *passim*, 71, 100 n.4, 234
FRG (Federal Republic of Germany) 10–17 *passim*, 47, 48, 54, 59–63 *passim*, 67–76 *passim*, 81, 82, 93, 104, 159, 189, 222, 224, 229, 230

General Electric 123
General Motors 67
Germany, *see* FRG
Gildemeister 93, 224

government policy 238–9, *see also under separate countries*
　European 226–32
　subsidizing 226, 228, 232
　NICs 193–284, *see also under separate countries*
　import prohibition 215, 219
　import restrictions 216, 219, 233
　protecting duties (tariffs) 194, 197–8, 199, 200–1, 208, 213, 214
　subsidizing 195–6, 200, 214–15, 220, 223, 232

hardware 70, 75
Heyligenstadt 189, 191
Holland 115
Hyundai Automobile 188, 190, 191

India 18, 30, 102–3
infant industries, protection of 194–5, 198–9, 201, 205, 211, 213
Italy (Italian) 11, 13, 17, 59, 62, 67, 71, 81, 99, 128, 139, 234

Japan(ese) 2, 4, 7, 10–13 *passim*, 16, 17, 31–7 *passim*, 52–67 *passim*, 72–91 *passim*, 95, 99, 101, 107, 134, 156, 161, 185, 210, 219–24 *passim*, 231
　CNC lathes
　　design of smaller 47 *et seq*
　　expansion in industry 31–7, 47–66
　　finance 62–3
　　pricing, competitive 50
　　sales, over the counter 50
　　simplification, emphasis on 49
　　smaller firms aimed for 50
　　trade 60 (table 3.13)
　　USA market, penetration into 223

Korea(n) 1, 3, 4, 18, 25–30 *passim*, 36, 89, 95, 102–5 *passim*, 106 (table 5.4), 109, 111, 131, 174–92, 198
　Advanced Institute of Science and Technology (KAIST) 188, 218
　CNC lathes industry 182–92, 210
　　import prohibition 219
　　import restrictions 219, 220
　exports
　　incentives 179–80, 191
　　machine tools 203 (table 9.2)
　　performance 203 (table 9.2)
　government policy 189, 190–1, 201–34
　import restrictions 177, 190–2 *passim*, 216, 219, 220

INDEX

Korea(n) – *contd.*
  imports and exports, share of 174 (table 8.1)
  long-term loans 181, 205–6
    National Investment Fund 206
  Machine tools 177, 180–1, 190
    exports 203 (table 9.2)
    government funds for local production 182, 185, 205
    protection of (tariffs) 180, 182
  subsidies, production 214–34

LAFTA (Latin-American Free Trade Organization) 117
lathes
  automatic 15, 17, (table 2.8), 18, 70
  choice of 12–18
  CNC (Computer numerical control), *see also* CNC 14–18, 69, 70, 91 (table 4.1)
    factor saving bias of 18–27
    prices in Japan, FRG and USA, comparison 10 (table 2.3)
  conventional 102
  copying 130
  cost(s) 13–15, 19–21
    investment 20–3
  engine 13, 17 (table 2.8), 18, 22, 91 (table 4.1), 150 (table 7.10), 156, 159, 161, 162, 185, 203, 209–11 *passim*
  flat-bed 107
  production of 16, 17, 203, 204–5
    CNC in 5 NICs 103
    CNC in Argentine, Taiwan and Korea 204 (table 9.5)
  revolver 15, 17, 18
  slant-bed 107
Leblond 81

machines 7, 7 (table 2.2)
machine tools, *see also* NCMT 6, 7, 8, 81, 92, 97, 111, 115, 141, 160–3 *passim*, 197, 201, 227
  Japanese 62
  NICs 102
machining centres 67–8, 95, 109, 155, 156, 159, 185–6
marketing 95–9, 104 (table 5.3)
  advertising 96
  after-sales service 95–7, 101 n.6
  network 108
  service/sales centre 108
  trade fairs 96

Marx, Karl 6
Massachusetts Institute of Technology 9
Maudslay, Henry 6
Mexico 30
micro chip 9
micro/computers/processors 9, 10, 28–9, 31, 55, 57, 72, 86 n.1, 90, 107
Mitsubishi 70
mobility between groups, barriers to 41
Mori Seiki 100

NCMT (Numerically Controlled Machine Tools) 9, 44–7, 52–66, 72, 96, 105, 108, 156, 166, 171
NIC(s) (Newly Industrializing Countries), *see also under separate countries* 1–4 *passim*, 18, 23–30 *passim*, 36, 79, 83, 86, 86 n.6, 91 (table 4.1), 93, 94, 100, 102–92, 231, 232, 237

OECD (Organization for Economic Co-operation and Development) 12, 13 (table 2.5), 17 (table 2.7), 37, 47, 68, 81–7, 86 n.6, 195
Olivetti 67–8, 86, 99

p-factor 18–19, 22–7 *passim*
Pontiggia PPL 86
product development 46, 74, 82
product/market combination, search for 41–2
  profitability, estimate of 42
profits 99

RAM (random access memory) 9
R & D (research and development) 39, 42, 50, 62, 72, 73, 83, 84, 88–90, 103–4, 133, 159, 168–72 *passim*, 185, 186, 217 (table 9.7), 218
retrofit 105, 107
robot(s) 12, 18, 69, 77, 224–5, 229
ROM (read only memory) 9

Sabatini Law 71
semiconductors 9
servo-techniques 90
Siemens 58, 70, 86 n.4, 123
Singapore 30
software 9, 70, 75–8 *passim*, 86, 91, 228–30 *passim*
South Africa 34
stocks, reduction of 22, 23

Sweden 2–5 *passim*, 13, 16–25 *passim*, 29, 43–8 *passim*, 54, 63, 64, 71–7, 79, 82, 85, 104, 115, 172, 221, 226–34 *passim*
system approach/design/development 70, 74–9 *passim*, 83, 110, 228, 231

Taiwan(ese) 1, 4, 18, 23–30 *passim*, 36, 89, 96, 102–6 *passim*, 106 (table 5.4), 111, 131, 143–73
  banking facilities 164–6, 169–70, 218
  CNC lathes 153–63, 210
  trade restrictions 219
  exported share 144 (table 7.3), 146, 203 (table 9.2)
  government policy 150–2, 163–8, 173, 193–220 *passim*
    subsidies 234
    tariffs 234
  import share 145 (tables 7.4 and 7.5), 146

Taiwan(ese) – *contd*.
  machine tool industry 146–50, 160–3
  U.S. distributors 148–9, 158
  MIRL (Mechanical Industry Research Laboratories) 166–7, 171–2, 173, 209–10, 218
trade unions 20
transfer lines 8

United Kingdom (UK) 2, 3, 5, 7, 11, 13, 16, 17, 24, 29, 43, 46, 55, 58, 59, 77, 82, 95, 161, 226–34 *passim*
United States of America (U.S.) 7, 9, 11, 13, 31–7 *passim*, 47, 48, 52–3, 59, 62, 63, 75, 79–83 *passim*, 96, 148, 149, 158, 161–3, 221–3

Volvo 45

Warner Swazy 81